Karatsuba   Basic Analytic Number Theory

Anatolij A. Karatsuba

Basic Analytic
Number Theory

Translated from the Russian
by Melvyn B. Nathanson

Springer-Verlag Berlin Heidelberg GmbH

Anatolij A. Karatsuba
Steklov Mathematical Institute, ul. Vavilova 42
117966, Moscow, Russia

Melvyn B. Nathanson
School of Mathematics, Institute for Advanced Study
Princeton, NJ 08540, and
Department of Mathematics, Lehman College (CUNY)
Bronx, NY 10468, USA

Title of the Russian edition:
Osnovy analiticheskoj teorii chisel, 2nd edition,
Publisher Nauka, Moscow 1983

Mathematics Subject Classification (1991):
11-01, 11M06, 11N05, 11N13, 11P05, 11P21, 11P32, 11P55

ISBN 978-3-540-53345-0

Library of Congress Cataloging-in-Publication Data
Karatsuba, Anatolii Alekseevich.
[Osnovy analiticheskoĭ teorii chisel. English]
Basic analytic number theory/Anatolij A. Karatsuba;
translated from the Russian by Melvyn B. Nathanson.
p. cm. Translation of: Osnovy analiticheskoĭ teorii chisel.
Includes bibliographical references and index.
ISBN 978-3-540-53345-0   ISBN 978-3-642-58018-5 (eBook)

DOI 10.1007/978-3-642-58018-5

1. Number theory. I. Title.
QA241.K3313 1993   512'.73 —dc20   91-40715

Typesetting: Macmillan India Limited, Bangalore 560 025
41/3140-543210- Printed on acid-free paper

Contents

Translator's Preface

This English translation of Karatsuba's *Basic Analytic Number Theory* follows closely the second Russian edition, published in Moscow in 1983. For the English edition, the author has considerably rewritten Chapter I, and has corrected various typographical and other minor errors throughout the the text.

August, 1991   Melvyn B. Nathanson

Introduction to the English Edition

It gives me great pleasure that Springer-Verlag is publishing an English translation of my book. In the Soviet Union, the primary purpose of this monograph was to introduce mathematicians to the basic results and methods of analytic number theory, but the book has also been increasingly used as a textbook by graduate students in many different fields of mathematics. I hope that the English edition will be used in the same ways. I express my deep gratitude to Professor Melvyn B. Nathanson for his excellent translation and for much assistance in correcting errors in the original text.

A.A. Karatsuba

Introduction to the Second Russian Edition

Number theory is the study of the properties of the integers. Analytic number theory is that part of number theory in which, besides purely number theoretic arguments, the methods of mathematical analysis play an essential role.

The purpose of this book is to provide an introduction to the central problems of analytic number theory. I have concentrated on the most important results, and have ignored excessively technical improvements. Consequently, I do not present the most recent refinements of the fundamental theorems. These latest results do not differ in principle from those that are presented in this book.

The book focuses on four problems in analytic number theory: the problem of integer points in planar domains, the problem of the distribution of prime numbers in the sequence of all natural numbers and in arithmetic progressions, Goldbach's problem, and Waring's problem. To solve these problems one must use the fundamental methods of analytic number theory: I.M. Vinogradov's method of trigonometric sums, the method of complex integration, and the circle method of G.H. Hardy, J.E. Littlewood, and S. Ramanujan.

There are exercises at the end of each chapter. They are related to the contents of the chapter, and the reader should try to solve them. The problems either refine the theorems proved in the text, or lead to new ideas in number theory.

The mathematical prerequisites for this book are undergraduate courses in number theory, mathematical analysis, and the theory of functions of a complex variable.

This book should be read consecutively, since each chapter is connected with its predecessors. If a topic occurs several times in the book, then it will be explained in detail only at its first appearance. The reader should understand and be able to justify every mathematical argument, line by line. Only in this way will the study of the book be useful.

The exercises in this book play a special role. Some of them are often very difficult, and can serve as topics for further research.

The books [1]–[12] listed in the Bibliography discuss various mathematical topics related to this volume, and also contain historical and other references.

Statements and formulas are numbered separately in each chapter. To refer to statements in other chapters, we note the chapter.

This second edition is considerably different from the first. I have added new material to Chapter I on integer points, as well as hints for the solution of most of the exercises in the book. There are also new proofs for various theorems in Chapters III–VII, X, and XI.

I received considerable assistance with this book from G.I. Arkhipov, S.M. Voronin, A.F. Lavrik, and V.N. Chubarikov. The manuscript was typed and edited by L.N. Abramochkina and R.I. Sorokina. I am very grateful to them all.

A.A. Karatsuba

Notation

$c, c_0, c_1, \ldots$ denote absolute positive constants, which are, in general, different in different theorems.

For positive A, the notations $B = O(A)$ and $B \ll A$ mean that $|B| \leq cA$; the notation $A \asymp B$ means that

$$c_1 A \leq B \leq c_2 A.$$

$\varepsilon, \varepsilon_1$ are arbitrarily small positive constants, and n, m, k, l, N denote natural numbers. Except in Chapter II, $p, p_1, \ldots$ denote prime numbers.

The Möbius function $\mu(n)$ is defined by

$$\mu(n) = \begin{cases} 1, & \text{if } n = 1; \\ 0, & \text{if } n = p^2 m; \\ (-1)^k, & \text{if } n = p_1 \ldots p_k \end{cases}$$

For $x > 0$, the logarithm and the logarithmic integral are defined by

$$\ln x = \log x = \int_1^x \frac{du}{u}; \quad \mathrm{Li}\, x = \int_2^x \frac{du}{\ln u} + c_0,$$

where

$$c_0 = \lim_{\delta \to 0} \left(\int_0^{1-\delta} \frac{du}{\ln u} + \int_{1+\delta}^2 \frac{du}{\ln u} \right); \quad \exp F = e^F.$$

The von Mangoldt function $\Lambda(n)$ is defined by

$$\Lambda(n) = \begin{cases} \log p & \text{if } n = p^k; \\ 0 & \text{if } n \neq p^k. \end{cases}$$

The Euler function $\varphi(k)$ is the number of natural numbers less than and relatively prime to k.

Chebyshev's function $\psi(x)$ and the prime counting function $\pi(x)$ are defined by

$$\psi(x) = \sum_{n \leq x} \Lambda(n); \quad \pi(x) = \sum_{p \leq x} 1.$$

For $l \leq k, (l, k) = 1$,

$$\psi(x; k, l) = \sum_{\substack{n = l \,(\mathrm{mod}\, k) \\ n \leq x}} \Lambda(n); \quad \pi(x; k, l) = \sum_{\substack{p = l \,(\mathrm{mod}\, k) \\ p \leq x}} 1.$$

The divisor function $\tau(n)$ is the number of positive divisors of n; $\tau_k(n)$ denotes the number of solutions of the equation $x_1 x_2 \ldots x_k = n$ in natural numbers $x_1, \ldots, x_k$. Thus, $\tau_2(n) = \tau(n)$.

The number of prime divisors of n is denoted by $\Omega(n)$.

For any real number α, the *integer part* of α, denoted by $[\alpha]$, is the largest integer not exceeding α; the *fractional part* of α, denoted by $\{\alpha\}$, is $\alpha - [\alpha]$; and $\|\alpha\| = \min(\{\alpha\}, 1 - \{\alpha\})$ is the distance from α to the nearest integer.

$s = \sigma + it$ denotes a complex number, where $i^2 = -1$, Re $s = \sigma$, Im $s = t$, $\bar{s} = \sigma - it$. In general, $\bar{f}$ is the complex conjugate of f. Everywhere, $\ln s = \log s$ is the principal branch of the logarithm, and γ is Euler's constant, defined by

$$\gamma = \lim_{m \to +\infty} \left(1 + \frac{1}{2} + \ldots + \frac{1}{m} - \ln m \right).$$

The nontrivial zeros of the Riemann-zeta function and the Dirichlet L- series are enumerated in order of the increasing absolute value of their imaginary parts. A trigonometric sum is a finite sum of the form

$$\sum_{n \leq P} G(n) \exp(2\pi i F(n)),$$

where G and F are real-valued functions of a natural number n, and

$$\exp(i\varphi) = \cos \varphi + i \sin \varphi.$$

Chapter I. Integer Points

In this chapter we consider two fundamental problems in the theory of integer points: Gauss's problem on the number of integer points inside a circle, and the Dirichlet divisor problem. We shall assume that a Cartesian coordinate (x, y) system has been defined on the plane.

§1. Statement of the Problem, Auxiliary Remarks, and the Simplest Results

Definition. *The point M with coordinates (x, y) is called an integer point if the numbers x and y are integers.*

We consider the circle $x^2 + y^2 \leq R$, and we denote by $K(R)$ the number of integer points inside this circle. For large R the value of $K(R)$ is approximately πR, the area of the circle. We denote by $\Delta(R)$ the difference between $K(R)$ and πR, that is,

$$\Delta(R) = K(R) - \pi R.$$

Gauss's problem on the number of integer points inside a circle is to determine the correct order of magnitude of $\Delta(R)$ as $R \to \infty$.

Dirichlet's divisor problem is similar. We consider the number $L(R)$ of integer points with positive coordinates under the hyperbola $xy = R$. Let

$$\Delta_1(R) = L(R) - R(\ln R + 2\gamma - 1), \quad R \to +\infty,$$

where γ is Euler's constant. We want to determine the correct order of magnitude of $\Delta_1(R)$ as $R \to \infty$.

Let $\tau(n)$ denote the number of divisors of n. It follows from the definitions of $L(R)$ and $\tau(n)$ that

$$L(R) = \sum_{n \leq R} \tau(n).$$

This explains the name "divisor problem."

These problems are special cases of the general problem of determining the number of integer points inside domains bounded by a curve $y = f(x)$, where $f(x)$ is a function that is continuous and nonnegative on the interval $[a, b]$, and by the lines $x = a$, $x = b$, and $y = 0$. We want to count the number of points $M = (x, y)$ satisfying the conditions $a < x \leq b$ and $0 < y \leq f(x)$. We denote this number by the letter T. Then

$$T = \sum_{a<x\leq b} [f(x)] = \sum_{a<x\leq b} f(x) - \sum_{a<x\leq b} \{f(x)\}. \tag{1}$$

This leads to two problems: (1) The calculation of the exact value of the first sum, and (2) the determination of a good asymptotic estimate for the second sum. Theorem 1 gives a solution to the first problem under certain general assumptions about the function $f(x)$. The second problem is the fundamentally difficult problem in the theory of integer points.

Theorem 1 (Euler-Maclaurin sum formula). *Let $f(x)$ be a twice continuously differentiable function on the interval $[a, b]$. Define the functions $\rho(x)$ and $\sigma(x)$ by the equations*

$$\rho(x) = \tfrac{1}{2} - \{x\}, \quad \sigma(x) = \int_0^x \rho(u)\, du.$$

Then

$$\sum_{a<x\leq b} f(x) = \int_a^b f(x)\, dx + \rho(b) f(b) - \rho(a) f(a) + \sigma(a) f'(a)$$

$$- \sigma(b) f'(b) + \int_a^b \sigma(x) f''(x)\, dx.$$

Proof. We shall assume that the interval $(a, b]$ contains at least one integer point. Partitioning the interval of integration at the integer points, we obtain

$$\int_a^b \sigma(x) f''(x)\, dx = \int_a^{[a]+1} \sigma(x) f''(x)\, dx + \sum_{n=[a]+1}^{[b]-1} \int_n^{n+1} \sigma(x) f''(x)\, dx$$

$$+ \int_{[b]}^b \sigma(x) f''(x)\, dx.$$

In each of these intervals of integration the functions $\rho(x)$ and $\sigma(x)$ are continuously differentiable, and $\sigma'(x) = \rho(x)$. Therefore, integrating by parts twice, we obtain

$$\int_a^{[a]+1} \sigma(x) f''(x)\, dx = -\sigma(a) f'(a) + \tfrac{1}{2} f([a] + 1) + \rho(a) f(a) - \int_a^{[a]+1} f(x)\, dx;$$

$$\int_n^{n+1} \sigma(x) f''(x)\, dx = \tfrac{1}{2} f(n + 1) + \tfrac{1}{2} f(n) - \int_n^{n+1} f(x)\, dx;$$

$$\int_{[b]}^b \sigma(x) f''(x)\, dx = \sigma(b) f'(b) - \rho(b) f(b) + \tfrac{1}{2} f([b]) - \int_{[b]}^b f(x)\, dx.$$

The theorem follows from substituting these identities into the preceding formula. If the interval $(a, b]$ contains no integer point, then the left side of the

Euler-Maclaurin formula equals 0. Integration by parts twice shows that the right side of the formula also equals 0. This completes the proof of the Theorem.

□

Remark. In applications it is often sufficient to use a simpler summation formula:

$$\sum_{a<x\leq b} f(x) = \int_a^b f(x)dx + \rho(b)f(b) - \rho(a)f(a) - \int_a^b \rho(x)f'(x)dx. \qquad (2)$$

This formula requires only that $f(x)$ be continuously differentiable on $[a, b]$. We proceed to the problems of Gauss and Dirichlet.

Theorem 2 (Gauss). *The following asymptotic formula holds for* $K(R)$:

$$K(R) = \pi R + \Delta(R)$$

$$\Delta(R) = O(\sqrt{R}).$$

Proof. We consider the curvilinear trapezoid

$$0 < x \leq \sqrt{R/2}, \quad 0 < y \leq \sqrt{R - x^2}$$

(cf. Fig. 1). The disc K: $x^2 + y^2 \leq R$ consists of eight such trapezoids. Since the intersections of these domains are squares with sides $\sqrt{R/2}$, it follows from formula 1 that

$$K(R) = 1 + 4[\sqrt{R}] + 8 \sum_{0<x\leq\sqrt{R/2}} [\sqrt{R - x^2}] - 4([\sqrt{R/2}])^2$$

$$= 8 \sum_{0<x\leq\sqrt{R/2}} \sqrt{R - x^2} - 2R + 4\sqrt{2R}\{\sqrt{R/2}\}$$

$$+ 4\sqrt{R} - 8 \sum_{0<x\leq\sqrt{R/2}} \{\sqrt{R - x^2}\} + O(1).$$

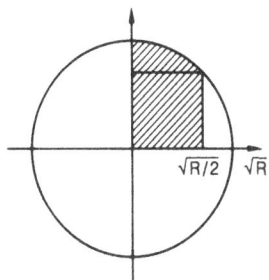

Fig. 1

We use Theorem 1 to compute the next to last sum.

$$
\sum_{0<x\le\sqrt{R/2}}\sqrt{R-x^2}=\int_0^{\sqrt{R/2}}\sqrt{R-u^2}\,du+\left(\frac{1}{2}-\left\{\sqrt{\frac{R}{2}}\right\}\right)\sqrt{\frac{R}{2}}
$$

$$
-\frac{\sqrt{R}}{2}+\sigma(\sqrt{R/2})\frac{\sqrt{R/2}}{\sqrt{R/2}}+\int_0^{\sqrt{R/2}}\sigma(u)\frac{d^2}{du^2}\left(\sqrt{R-u^2}\right)du.
$$

Since $|\sigma(u)|\le 1/8$ and since the first integral is equal to the area of the curvilinear trapezoid, which is $\pi R/8 + R/4$, it follows that

$$
\sum_{0<x\le\sqrt{R/2}}\sqrt{R-x^2}=\frac{\pi R}{8}+\frac{R}{4}+\frac{\sqrt{R}}{2\sqrt{2}}-\sqrt{\frac{R}{2}}\left\{\sqrt{\frac{R}{2}}\right\}-\frac{\sqrt{R}}{2}+O(1).
$$

This implies that

$$
K(R)=\pi R+\Delta(R),
$$

where

$$
\Delta(R)=2\sqrt{2R}-8\sum_{0<x\le\sqrt{R/2}}\{\sqrt{R-x^2}\}+O(1)=O(\sqrt{R}),\qquad(3)
$$

which is what we had to prove.   □

Theorem 3 (Dirichlet). *The following asymptotic formula holds for $L(R)$:*

$$
L(R)=R(\ln R+2\gamma-1)+\Delta_1(R),\quad \Delta_1(R)=O(\sqrt{R}),
$$

where γ is Euler's constant.

Proof. Consider the curvilinear trapezoid (cf. Fig. 2) $1\le x\le\sqrt{R}$, $0<y\le R/x$. The domain L is the union of two domains, each equal to this trapezoid. Applying formula 1, we find that

$$
L(R)=2\sum_{1\le x\le\sqrt{R}}\left[\frac{R}{x}\right]-([\sqrt{R}])^2
$$

$$
=2R\sum_{1\le x\le\sqrt{R}}\frac{1}{x}-2\sum_{1\le x\le\sqrt{R}}\left\{\frac{R}{x}\right\}-R+2\sqrt{R}\{\sqrt{R}\}+O(1).
$$

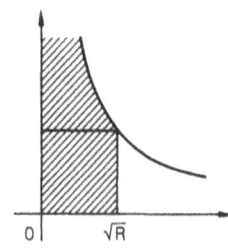

Fig. 2

It follows from formula 1 that

$$\sum_{1 < x \leq \sqrt{R}} \frac{1}{x} = \int_1^{\sqrt{R}} \frac{du}{u} + \left(\frac{1}{2} - \{\sqrt{R}\}\right) \frac{1}{\sqrt{R}} - \frac{1}{2} + \sigma(\sqrt{R})\frac{1}{R} - \sigma(1) + 2\int_1^{\sqrt{R}} \sigma(u) \frac{du}{u^3}$$

$$= \ln\sqrt{R} + \left(\frac{1}{2} - \{\sqrt{R}\}\right)\frac{1}{\sqrt{R}} - \frac{1}{2} + 2\int_1^{\infty} \sigma(u) \frac{du}{u^3} + O\left(\frac{1}{R}\right).$$

Euler's constant is defined by

$$\gamma = \lim_{Y \to +\infty} \left(\sum_{1 \leq n \leq Y} \frac{1}{n} - \ln Y\right).$$

Applying Theorem 1 to the sum in the parentheses, we find that

$$\gamma = -\frac{1}{2} + 2\int_1^{\infty} \sigma(u) \frac{du}{u^3}.$$

In the same way we obtain a formula for the value of $L(R)$:

$$L(R) = R\ln R + 2\sqrt{R}\left(\frac{1}{2} - \{\sqrt{R}\}\right) + (2\gamma - 1)R$$

$$- 2\sum_{1 \leq x \leq \sqrt{R}} \left\{\frac{R}{x}\right\} + 2\sqrt{R}\{\sqrt{R}\} + O(1)$$

$$= R\,(\ln R + 2\gamma - 1) + \Delta_1(R),$$

where

$$\Delta_1(R) = \sqrt{R} - 2\sum_{1 \leq x \leq \sqrt{R}} \left\{\frac{R}{x}\right\} + O(1) = O(\sqrt{R}), \tag{4}$$

which is what had to be proved.   □

Remark. If the fractional parts of the functions $f(x) = \sqrt{R - x^2}$ for $0 < x \leq \sqrt{R/2}$ and $f(x) = R/x$ for $0 < x \leq \sqrt{R}$ are uniformly distributed, i.e. if the proportion of fractional parts of $f(x)$ that belong to the interval $(a, b) \subseteq [0, 1]$ is equal to the length of (a, b), then

$$\sum_{0 < x \leq \sqrt{R/2}} \{\sqrt{R - x^2}\} = \frac{1}{2}\sqrt{\frac{R}{2}} + o(\sqrt{R}),$$

$$\sum_{0 < x \leq \sqrt{R}} \left\{\frac{R}{x}\right\} = \frac{1}{2}\sqrt{R} + o(\sqrt{R})$$

To obtain these estimates, it suffices to decompose $[0, 1)$ into "small" equal sub-intervals. These estimates give the following refinements of Theorems 2 and 3:

$$\Delta(R) = o(\sqrt{R}), \quad \Delta_1(R) = o(\sqrt{R}).$$

§2. The Connection Between Problems in the Theory of Integer Points and Trigonometric Sums

The probem of integer points in §1 reduces to the study of the asymptotic behavior of sums of the fractional parts of $f(x)$. This is closely connected with the problem of the distribution of the values of $\{f(x)\}$, which in turn leads to the investigation of trigonometric sums. The establishment of these connections is the subject of §2.

In the study of these problems, it is often useful to consider certain functions that are similar to the characteristic functions of intervals, but that are significantly more smooth.

Such functions are constructed in the following Lemma.

Lemma A. *Let $r \geq 1$ be an integer, and let α, β, and Δ be real numbers such that $0 < \Delta < 1/4$ and $\Delta \leq \beta - \alpha \leq 1 - \Delta$. Then there exists a function $\psi(x)$ that is periodic with period 1 and satisfies the conditions*

(1) $\psi(x) = 1$ *on the interval* $\alpha + \dfrac{\Delta}{2} \leq x \leq \beta - \dfrac{\Delta}{2}$;

(2) $0 < \psi(x) < 1$ *on the intervals*

$$\alpha - \frac{\Delta}{2} < x < \alpha + \frac{\Delta}{2} \quad \text{and} \quad \beta - \frac{\Delta}{2} < x < \beta + \frac{\Delta}{2};$$

(3) $\psi(x) = 0$ *on the interval* $\beta + \dfrac{\Delta}{2} \leq x \leq 1 + \alpha - \dfrac{\Delta}{2}$;

(4) *The Fourier expansion of $\psi(x)$ is of the form*

$$\psi(x) = \beta - \alpha + \sum_{\substack{m = -\infty \\ m \neq 0}}^{+\infty} g(m)e^{2\pi imx},$$

where

$$|g(m)| \leq \min\left(\beta - \alpha, \ \frac{1}{\pi|m|}, \ \frac{1}{\pi|m|}\left(\frac{r}{\pi|m|\Delta}\right)^r\right).$$

Proof. We define the function $\psi_0(x)$ by the equations (cf. Fig. 3)

1) $\psi_0(x) = 1$ for $\alpha < x < \beta$;
2) $\psi_0(\alpha) = \psi_0(\beta) = 1/2$;
3) $\psi_0(x) = 0$ for $\beta < x < 1 + \alpha$;
4) $\psi_0(x) = \psi_0(x + 1)$.

Decomposing this into a Fourier series, we find that

$$\psi_0(x) = a_{0,0} + \sum_{\substack{m = -\infty \\ m \neq 0}}^{+\infty} a_{m,0} e^{2\pi imx},$$

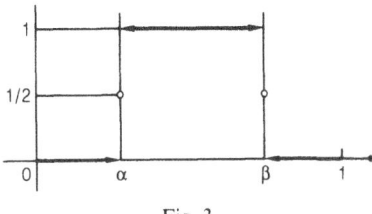

Fig. 3

where

$$a_{0,0} = \int_0^1 \psi_0(x)dx = \beta - \alpha;$$

$$a_{m,0} = \int_0^1 \psi_0(x)e^{-2\pi imx}dx$$

$$= \int_\alpha^\beta e^{-2\pi imx}dx = i\frac{e^{-2\pi im\beta} - e^{-2\pi im\alpha}}{2\pi m} \qquad \text{for } m \neq 0.$$

Now let $\delta = \Delta/(2r)$ and define successively the functions $\psi_1(x), \ldots, \psi_\rho(x)$ by the equations

$$\psi_\rho(x) = \frac{1}{2\delta}\int_{-\delta}^{+\delta}\psi_{\rho-1}(x+u)du.$$

The functions $\psi_\rho(x)$ have the following properties:

1) $\psi_\rho(x) = 1$ for $\alpha + \rho\delta \leq x \leq \beta - \rho\delta$;

2) $0 < \psi_\rho(x) < 1$ for $\alpha - \rho\delta < x < \alpha + \rho\delta$ and for $\beta - \rho\delta < x < \beta + \rho\delta$;

3) $\psi_\rho(x) = 0$ for $\beta + \rho\delta \leq x \leq 1 + \alpha - \rho\delta$;

4) $\psi_\rho(x) = \beta - \alpha + \sum_{\substack{m=-\infty \\ m \neq 0}}^{+\infty} a_{m,\rho}e^{2\pi imx},$

where

$$a_{m,\rho} = i\frac{e^{-2\pi im\beta} - e^{-2\pi im\alpha}}{2\pi m}\left(\frac{e^{2\pi im\delta} - e^{-2\pi im\delta}}{4\pi im\delta}\right)^\rho.$$

Properties 1–3 follow immediately from the definition of $\psi_\rho(x)$ and the properties of $\psi_{\rho-1}(x)$. We shall prove property 4 on the assumption that it holds for $\psi_{\rho-1}(x)$.

We have

$$a_{0,\rho} = \int_0^1 \psi_\rho(x)dx = \int_0^1 \frac{1}{2\delta}\left(\int_{-\delta}^{+\delta}\psi_{\rho-1}(x+u)du\right)dx$$

$$= \frac{1}{2\delta}\int_{-\delta}^{+\delta}du\left(\int_0^1 \psi_{\rho-1}(x+u)dx\right) = \beta - \alpha;$$

$$a_{m,\rho} = \int_0^1 \psi_\rho(x)e^{-2\pi imx}dx = \int_0^1 \frac{1}{2\delta}\left(\int_{-\delta}^{+\delta}\psi_{\rho-1}(x+u)\,du\right)e^{-2\pi imx}dx$$

$$= \frac{1}{2\delta}\int_{-\delta}^{+\delta} du\left(\int_0^1 \psi_{\rho-1}(x+u)e^{-2\pi imx}dx\right)$$

$$= \frac{1}{2\delta}\int_{-\delta}^{+\delta} e^{2\pi imu}\,du\left(\int_0^1 \psi_{\rho-1}(x+u)e^{-2\pi im(x+u)}dx\right)$$

$$= \frac{a_{m,\rho-1}}{2\delta}\cdot\frac{e^{2\pi im\delta}-e^{-2\pi im\delta}}{2\pi im}$$

$$= i\frac{e^{-2\pi im\beta}-e^{-2\pi im\alpha}}{2\pi m}\left(\frac{e^{2\pi im\delta}-e^{-2\pi im\delta}}{4\pi im\delta}\right)^\rho$$

for $m \ne 0$, which is what we had to prove. Setting $\rho = r$ completes the proof of the Lemma. □

Figure 4 is a drawing of the graph of the function $\psi(x)$.

The following lemma reduces the question of the asymptotic behaviour of the sums of the fractional parts of different functions to an estimate for trigonometric sums.

Lemma B. *Let* $\delta_1, \delta_2, \ldots, \delta_Q$ *be real numbers,* $0 \le \delta_s < 1$, $s = 1, 2, \ldots, Q$. *Moreover, let* $r \ge 1$ *be an integer, let* $0 < \Delta < 1/8$, *and let* R, α, *and* β *be real numbers satisfying the condition* $\Delta \le \beta - \alpha \le 1 - \Delta$. *Let* $\psi(x)$ *be the function constructed in Lemma A for the numbers* Δ, α, β, r. *Suppose that, for all permissable values of* α *and* β, *the sum* $U(\alpha, \beta) = \sum_{s=1}^{Q} \psi(\delta_s)$ *satisfies the relation* $U(\alpha, \beta) = (\beta - \alpha)Q + O(R)$. *Then*

(a) *For any* σ *satisfying* $0 < \sigma \le 1$, *the number* A_σ *of real numbers* δ_s *such that* $\delta_s < \sigma$ *can be expressed by the formula* $A_\sigma = \sigma Q + R_\sigma$, *where* $R_\sigma = O(R) + O(\Delta Q)$;

(b) *The following equation holds:*

$$S = \sum_{s=1}^{Q} \delta_s = \tfrac{1}{2}Q + O(R) + O(\Delta Q).$$

Proof. For $0 < \beta - \alpha \le 1$ we denote by $D(\alpha, \beta)$ the set of numbers δ_s such that $\alpha \le \delta_s < \beta \pmod 1$. If $2\Delta < \beta - \alpha \le 1 - 2\Delta$, then it follows from the obvious

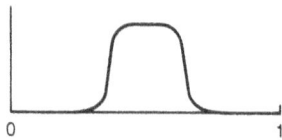

Fig. 4

inequality

$$U\left(\alpha + \frac{\Delta}{2}, \beta - \frac{\Delta}{2}\right) \le D(\alpha, \beta) \le U\left(\alpha - \frac{\Delta}{2}, \beta + \frac{\Delta}{2}\right)$$

and from the conditions of the Lemma that

$$D(\alpha, \beta) = (\beta - \alpha)Q + O(R) + O(\Delta Q).$$

This relation also holds for arbitrary α and β that satisfy the inequality $0 < \beta - \alpha \le 1$. This follows from the following identities:
If $0 < \beta - \alpha < 2\Delta$, then

$$D(\alpha, \beta) = D(\alpha, \alpha + 1 - 2\Delta) - D(\beta, \alpha + 1 - 2\Delta);$$

if $1 - 2\Delta \le \beta - \alpha < 1$, then

$$D(\alpha, \beta) = D(\alpha, \alpha + \tfrac{1}{2}) + D(\alpha + \tfrac{1}{2}, \beta).$$

Letting $\alpha = 0$ and $\beta = \sigma$, we obtain the first assertion of the Lemma.

(b) We shall assume that $R < 0.1Q$, since in the opposite case the Lemma holds trivially. We set $n = [QR^{-1}]$ and $v = 1/n$. According to (a), with $A_0 = 0$ and $R_0 = 0$, we find that

$$
\begin{aligned}
A_v &= vQ + R_v, & A_v - A_0 &= vQ + R_v - R_0, \\
A_{2v} &= 2vQ + R_{2v}, & A_{2v} - A_v &= vQ + R_{2v} - R_v, \\
A_{3v} &= 3vQ + R_{3v}, & A_{3v} - A_{2v} &= vQ + R_{3v} - R_{2v},
\end{aligned}
$$

. . . .

$$A_{nv} = nvQ + R_{nv}, \quad A_{nv} - A_{(n-1)v} = vQ + R_{nv} - R_{(n-1)v}.$$

The cardinality of the set of numbers δ satisfying the condition $(k-1)v \le \delta_s < kv$ equals $A_{kv} - A_{(k-1)v}$ for $k = 1, 2, \ldots, n$. Therefore, multiplying the expressions for $A_{kv} - A_{(k-1)v}$ by $(k-1)v$ and adding, we obtain a lower bound for S. Multiplying by kv and adding, we obtain an upper bound, i.e.

$$\sum_{k=1}^{n} (vQ + R_{kv} - R_{(k-1)v})(k-1)v \le S \le \sum_{k=1}^{n} (vQ + R_{kv} - R_{(k-1)v})kv.$$

Moreover,

$$\sum_{k=1}^{n} v^2 Q(k-1) = v^2 Q \frac{n(n-1)}{2} = \frac{Q}{2} + O(R);$$

$$\sum_{k=1}^{n} v^2 Qk = v^2 Q \frac{n(n+1)}{2} = \frac{Q}{2} + O(R);$$

$$\sum_{k=1}^{n} vk(R_{kv} - R_{(k-1)v}) = v \sum_{k=1}^{n} (kR_{kv} - (k-1)R_{(k-1)v} - R_{(k-1)v})$$

$$= v(nR_{nv} - R_{1v} - \ldots - R_{(n-1)v}) = O(R) + O(\Delta Q);$$

$$\sum_{k=1}^{n} (R_{kv} - R_{(k-1)v})(k - 1)v = O(R) + O(\Delta Q).$$

Statement (b) follows trivially from this. □

It follows from the Lemma B that, for the asymptotic estimation of a sum of the values of $\{f(x)\}$, we must be able to compute the sum of the values of $\psi(f(x))$. It follows from Lemma A that this computation reduces to estimates of trigonometric sums of the form

$$\sum_{x} e^{2\pi i m f(x)}, \quad m \neq 0.$$

§3. Theorems on Trigonometric Sums

Under the appropriate conditions on the function $f(x)$ in the exponent, a trigonometric sum can be transformed into another one that is "shorter," i.e. has fewer terms, and so the trivial estimate of the latter sum gives a nontrivial estimate for the original sum. We must first prove a lemma on partial summation.

Lemma C (partial summation). *Let $f(x)$ be a continuously differentiable function on $[a, b]$, let c_n be arbitrary complex numbers, and let*

$$C(x) = \sum_{a < n < x} c_n.$$

Then

$$\sum_{a < n \leq b} c_n f(n) = -\int_a^b C(x) f'(x) dx + C(b) f(b).$$

Proof. We have

$$C(b) f(b) - \sum_{a < n \leq b} c_n f(n) = \sum_{a < n \leq b} c_n (f(b) - f(n))$$

$$= \sum_{a < n \leq b} \int_n^b c_n f'(x) dx = \sum_{a < n \leq b} \int_a^b c_n g(n; x) f'(x) dx,$$

where

$$g(n; x) = \begin{cases} 1, & \text{if } n \leq x \leq b; \\ 0, & \text{if } x < n. \end{cases}$$

Changing the order of summation and integration, and observing that

$$\sum_{a < n \leq b} c_n g(n; x) = \sum_{a < n \leq x} c_n = C(x),$$

we complete the proof of the Lemma. □

Lemma 1. *Let $f(x)$ and $\phi(x)$ be real-valued functions satisfying the following conditons on the interval $[a, b]$:*

1. *$f''(x)$ and $\phi'(x)$ are continuous,*
2. *$0 < f''(x)$,*
3. *There exist numbers $H > 0$, and $U \geq b - a > 0$ such that $\phi(x) \ll H$, $\phi'(x) \ll HU^{-1}$.*

Then for any $\Delta \in (0, 1)$

$$\sum_{a < x \leq b} \varphi(x) e^{2\pi i f(x)} = \sum_{\alpha - \Delta \leq n \leq \beta + \Delta} \int_a^b \varphi(x) e^{2\pi i (f(x) - nx)} \, dx + O(H \ln(\beta - \alpha + 2)),$$

(5)

where $\alpha = f'(a)$, $\beta = f'(b)$, and the constant in the O symbol depends only on Δ.

Proof. We shall assume that $b - a > 10$; otherwise, the Lemma is trivial. We first let $\varphi(x) = 1$. Let $m = [10U^2 K]$ and $K = 1 + |\alpha| + |\beta|$. For $[a] + 2 \leq M \leq [b] - 1$, we consider the integral

$$W_M = \int_{-0.5}^{+0.5} \frac{\sin(2m + 1)\pi x}{\sin \pi x} e^{2\pi i f(M + x)} dx.$$

Since

$$\frac{\sin(2m + 1)\pi x}{\sin \pi x} = \sum_{n = -m}^{m} e^{2\pi i n x},$$

(6)

it follows that

$$\int_{-0.5}^{+0.5} \frac{\sin(2m + 1)\pi x}{\sin \pi x} dx = 1,$$

and so

$$W_M = e^{2\pi i f(M)} + V_M,$$

where

$$V_M = \int_{-0.5}^{+0.5} \frac{\sin(2m + 1)\pi x}{\sin \pi x} (e^{2\pi i f(M + x)} - e^{2\pi i f(M)}) dx.$$

In order to estimate $|V_M|$ from above, we represent V_M as a sum of three integrals:

$$V_M = \int_{-\frac{1}{m}}^{\frac{1}{m}} + \int_{-\frac{1}{2}}^{-\frac{1}{m}} + \int_{\frac{1}{m}}^{\frac{1}{2}}.$$

We estimate the first integral by applying the finite difference formula to the difference inside the parentheses:

$$\int_{-\frac{1}{m}}^{\frac{1}{m}} \ll \int_{-\frac{1}{m}}^{\frac{1}{m}} \frac{|f'| |x| \, dx}{|x|} \ll \frac{K}{m} \ll U^{-1}.$$

We can estimate the second and third integrals in the same way. Thus, to estimate the second, we first integrate it by parts and obtain

$$
\int\limits_{\frac{1}{m}}^{\frac{1}{2}} \frac{\sin(2m+1)\pi x}{\sin \pi x} \left(e^{2\pi i f(M+x)} - e^{2\pi i f(M)} \right) dx
$$

$$
= - \frac{e^{2\pi i f(M+x)} - e^{2\pi i f(M)}}{\sin \pi x} \cdot \frac{\cos(2m+1)\pi x}{(2m+1)\pi} \Big|_{\frac{1}{m}}^{\frac{1}{2}} + \int\limits_{\frac{1}{m}}^{\frac{1}{2}} \frac{\cos(2m+1)\pi x}{(2m+1)\pi} Y_x dx,
$$

where

$$
Y_x = \frac{e^{2\pi i f(M+x)} 2\pi i f'(M+x)}{\sin \pi x} - \frac{(e^{2\pi i f(M+x)} - e^{2\pi i f(M)})\pi \cos \pi x}{\sin^2 \pi x}
$$

The first summand has order of magnitude

$$
O\left(\frac{K}{m}\right) = O(U^{-1}).
$$

Further,

$$
Y_x \ll \frac{K}{|x|}, \quad \int\limits_{\frac{1}{m}}^{\frac{1}{2}} \frac{\cos(2m+1)\pi x}{(2m+1)\pi} Y_x dx \ll \frac{K}{m} \ln m \ll U^{-1},
$$

and so

$$
V_M \ll U^{-1}; \quad W_M = e^{2\pi i f(M)} + O(U^{-1}).
$$

Summing the last relation over M and using formula (6) and the definition of W_M, we obtain

$$
\sum_{a < x \le b} e^{2\pi i f(x)} = \sum_{M=[a]+2}^{[b]-1} W_M + O(1)
$$

$$
= \sum_{M=[a]+2}^{[b]-1} \int\limits_{-0.5}^{+0.5} \sum_{n=-m}^{m} e^{2\pi i (f(M+x)-nx)} dx + O(1)
$$

$$
= \sum_{n=-m}^{m} \sum_{M=[a]+2}^{[b]-1} \int\limits_{M-0.5}^{M+0.5} e^{2\pi i (f(x)-nx)} dx + O(1)
$$

$$
= \sum_{n=-m}^{m} I_n + O(1), \tag{7}
$$

where

$$
I_n = \int\limits_{[a]+1.5}^{[b]-0.5} e^{2\pi i (f(x)-nx)} dx.
$$

Next we estimate the sum in (7) for $n < f'(a) - \Delta$ and $n > f'(b) + \Delta$, i.e. the sum Σ, where

$$
\Sigma = \sum_{-m \le n < f'(a) - \Delta} I_n + \sum_{f'(b) + \Delta < n \le m} I_n.
$$

Integrating once by parts, we find

$$I_n = \frac{e^{2\pi i(f(x)-nx)}}{2\pi i(f'(x)-n)}\bigg|_{a_1}^{b_1} + O\left(\int_{a_1}^{b_1} \frac{f''(x)dx}{(f'(x)-n)^2}\right)$$

$$= \frac{e^{\pi in}}{2\pi i}\left(\frac{e^{2\pi if(b_1)}}{f'(b_1)-n} - \frac{e^{2\pi if(a_1)}}{f'(a_1)-n}\right) + O\left(\frac{f'(b_1)-f'(a_1)}{(f'(a_1)-n)(f'(b_1)-n)}\right),$$

where $a_1 = [a] + 1.5$ and $b_1 = [b] - 0.5$. Since the fractions

$$\frac{1}{f'(b_1)-n} \quad \text{and} \quad \frac{1}{f'(a_1)-n}$$

increase monotonically for $-m \le n < f'(a) - \Delta$ and $f'(b) + \Delta < n \le m$, and since $e^{\pi in} = (-1)^n$, it follows that

$$\sum \frac{(-1)^n}{f'(b_1)-n} = O(1), \qquad \sum \frac{(-1)^n}{f'(a_1)-n} = O(1).$$

Moreover, for $-m \le n < f'(a) - \Delta$, the fraction

$$\frac{f'(b_1)-f'(a_1)}{(f'(a_1)-n)(f'(b_1)-n)}$$

increases monotonically, and so

$$\sum_{-m \le n < f'(a)-\Delta} \frac{f'(b_1)-f'(a_1)}{(f'(a_1)-n)(f'(b_1)-n)} \ll \frac{f'(b_1)-f'(a_1)}{\Delta(f'(b_1)-f'(a_1)+\Delta)}$$

$$+ \int_{-m}^{f'(a)-\Delta} \frac{f'(b_1)-f'(a_1)}{(f'(a_1)-x)(f'(b_1)-x)}\,dx = O(\ln(\beta - \alpha + 2)).$$

In the same way we can estimate the sum of these fractions for $f'(b) + \Delta < n \le m$. Thus,

$$\Sigma = O(\ln(\beta - \alpha + 2));$$

$$\sum_{a < x \le b} e^{2\pi if(x)} = \sum_{\alpha - \Delta \le n \le \beta + \Delta} I_n + O(\ln(\beta - \alpha + 2)).$$

We can replace I_n by an integral over the interval (a, b):

$$I_n = \int_a^b e^{2\pi i(f(x)-nx)}\,dx - I_{n,1} - I_{n,2},$$

where

$$I_{n,1} = \int_a^{a_1} e^{2\pi i(f(x)-nx)}\,dx, \qquad I_{n,2} = \int_{b_1}^b e^{2\pi i(f(x)-nx)}\,dx.$$

Further, if $\|x\| \ne 0$, we find

$$\left|\sum_{\alpha - \Delta \le n \le \beta + \Delta} e^{-2\pi inx}\right| = \left|\frac{e^{-2\pi in_1x} - e^{-2\pi in_2x}}{e^{-2\pi ix} - 1}\right| \le \frac{1}{|\sin \pi x|} \le \frac{1}{2\|x\|}.$$

Therefore,

$$\sum_{\alpha-\Delta\le n\le\beta+\Delta}(I_{n,1}+I_{n,2})=O\left(\int_a^{a_1}\min\left(\beta-\alpha+2,\frac{1}{\|x\|}\right)dx\right)$$

$$+O\left(\int_{b_1}^b\min\left(\beta-\alpha+2,\frac{1}{\|x\|}\right)dx\right)=O(\ln(\beta-\alpha+2));$$

$$\sum_{a<x\le b}e^{2\pi i f(x)}=\sum_{\alpha-\Delta\le n\le\beta+\Delta}\int_a^b e^{2\pi i(f(x)-nx)}dx+O(\ln(\beta-\alpha+2)).$$

Now let $\varphi(x)$ be an arbitrary function. Applying Lemma C, we find

$$\sum_{a<x\le b}\varphi(x)e^{2\pi i f(x)}=-\int_a^b\mathbb{C}(u)\varphi'(u)du+\mathbb{C}(b)\varphi(b),$$

where

$$\mathbb{C}(u)=\sum_{a<x\le u}e^{2\pi i f(x)}.$$

Applying what has been proved above to $\mathbb{C}(u)$, we obtain

$$\mathbb{C}(u)=\sum_{\alpha-\Delta\le n\le\gamma+\Delta}\int_a^u e^{2\pi i(f(x)-nx)}dx+O(\ln(\beta-\alpha+2)),$$

$$\gamma=f'(u)\le\beta.$$

Since for $\gamma+\Delta<n\le\beta+\Delta$,

$$\int_a^u e^{2\pi i(f(x)-nx)}dx=O\left(\frac{1}{n-\gamma}\right),$$

it follows that

$$\mathbb{C}(u)=\sum_{\alpha-\Delta\le n\le\beta+\Delta}\int_a^u e^{2\pi i(f(x)-nx)}dx+O(\ln(\beta-\alpha+2));$$

$$\sum_{a<x\le b}\varphi(x)e^{2\pi i f(x)}=-\sum_{\alpha-\Delta\le n\le\beta+\Delta}\int_a^b\left(\int_a^u e^{2\pi i(f(x)-nx)}dx\right)\varphi'(u)du$$

$$+\sum_{\alpha-\Delta\le n\le\beta+\Delta}\left(\int_a^b e^{2\pi i(f(x)-nx)}dx\right)\varphi(b)+O(H\ln(\beta-\alpha+2))$$

$$=\sum_{\alpha-\Delta\le n\le\beta+\Delta}\int_a^b\varphi(x)e^{2\pi i(f(x)-nx)}dx+O(H\ln(\beta-\alpha+2)).$$

This completes the proof of the Lemma.   □

Corollary. *Assume that the conditions of Lemma 1 are satisfied, and assume, in addition, that $|f'(x)|\le\delta<1$ and $a\le x\le b$. Then*

$$\sum_{a<x\le b}\varphi(x)e^{2\pi i f(x)}=\int_a^b\varphi(x)e^{2\pi i f(x)}dx+O(H)$$

where the constant in the symbol O depends only on δ.

Lemma 2. *Let $f(x)$ and $\varphi(x)$ be real-valued functions that satisfy the following conditions on the interval $[a, b]$:*

1. $f^{(4)}(x)$ *and* $\varphi''(x)$ *are continuous,*

2. *there exist numbers* H, U, *and* A *with* $H > 0$, $U \gg A \gg 1$, *and* $U \geq b - a > 0$ *such that*

$$A^{-1} \ll f''(x) \ll A^{-1}, \quad f'''(x) \ll A^{-1}U^{-1}, \quad f^{(4)}(x) \ll A^{-1}U^{-2},$$

$$\varphi(x) \ll H, \quad \varphi'(x) \ll HU^{-1}, \quad \varphi''(x) \ll HU^{-2};$$

3. $f'(c) = 0$ *for some* $c \in [a, b]$.

Then the following formula holds:

$$\int_a^b \varphi(x)e^{2\pi i f(x)}dx = \frac{1+i}{\sqrt{2}} \frac{\varphi(c)e^{2\pi i f(c)}}{\sqrt{f''(c)}} + O(H)$$

$$+ O\left(H\min\left(\frac{1}{|f'(a)|}, \sqrt{A}\right)\right) + O\left(H\min\left(\frac{1}{|f'(b)|}, \sqrt{A}\right)\right).$$

Proof. Dividing the interval of integration into two parts at the point c, we obtain

$$\int_a^b \varphi(x)e^{2\pi i f(x)}dx = \int_a^c + \int_c^b. \tag{8}$$

Let us compute the second integral. It is clear that

$$\int_c^b \varphi(x)e^{2\pi i f(x)}dx = \varphi(c)\int_c^b e^{2\pi i f(x)}dx + \int_0^{b-c} (\varphi(x+c) - \varphi(c))e^{2\pi i f(x+c)}dx.$$

We estimate the second integral. Integration once by parts yields

$$\int_0^{b-c} (\varphi(x+c) - \varphi(c))e^{2\pi i f(x+c)}dx = \frac{\varphi(x+c) - \varphi(c)}{2\pi i f'(x+c)} e^{2\pi i f(x+c)}\bigg|_0^{b-c}$$

$$- \frac{1}{2\pi i}\int_0^{b-c} \frac{\varphi'(x+c)f'(x+c) - f''(x+c)(\varphi(x+c) - \varphi(c))}{(f'(x+c))^2} e^{2\pi i f(x+c)}dx.$$

Since

$$\varphi(x+c) = \varphi(c) = \varphi'(c)x + O(HU^{-1}x^2),$$
$$f'(x+c) = f''(c)x + O(A^{-1}U^{-1}x^2), \quad f'(x+c) \gg xA^{-1},$$
$$f''(x+c) = f''(c) + O(A^{-1}U^{-1}x),$$

then the first term has order of magnitude $O(H)$, and the numerator of the expression inside the integral has an absolute value that does not exceed

$$|(\varphi'(c) + O(HU^{-1}x))(f''(c)x + O(A^{-1}U^{-1}x^2)) - (f''(c)$$

$$+ O(A^{-1}U^{-1}x))(\varphi'(c)x + O(HU^{-1}x^2))| \ll HA^{-1}U^{-2}x^2.$$

Therefore, the entire integral has order of magnitude $O(H)$. Next we compare

J and J_1, where

$$J = \int_0^{b-c} e^{2\pi i f(x+c)} dx, \quad J_1 = \int_0^{b-c} \frac{f'(x+c)e^{2\pi i f(x+c)} dx}{\sqrt{2f''(c)(f(x+c)-f(c))}}.$$

Consider the difference $J - J_1$. Integration once by parts yields

$$J - J_1 = \frac{1}{2\pi i}\left(\frac{1}{f'(x+c)} - \frac{1}{\sqrt{2f''(c)(f(x+c)-f(c))}}\right)e^{2\pi i f(x+c)}\bigg|_0^{b-c}$$
$$+ O\left(\int_0^{b-c}\left|-\frac{f''(x+c)}{(f'(x+c))^2} + \frac{f'(x+c)}{\sqrt{8f''(c)(f(x+c)-f(c))^3}}\right| dx\right).$$

Applying the formulas

$$f(x+c) - f(c) = \tfrac{1}{2}f''(c)x^2 + \tfrac{1}{6}f'''(c)x^3 + O(A^{-1}U^{-2}x^4),$$
$$f'(x+c) = f''(c)x + \tfrac{1}{2}f'''(c)x^2 + O(A^{-1}U^{-2}x^3),$$
$$f''(x+c) = f''(c) + f'''(c)x + O(A^{-1}U^{-2}x^2),$$

we see that the first sum has order of magnitude $O(1)$, and the expression inside the integral of the second sum has order of magnitude $O(U^{-1})$. Consequently,

$$J = J_1 + O(1).$$

To compute J_1, we change the variable of integration by setting $f(x+c) - f(c) = u$. Denoting by λ the difference $f(b) - f(c)$, we have

$$J_1 = \frac{e^{2\pi i f(c)}}{\sqrt{2f''(c)}}\int_0^\lambda \frac{e^{2\pi i u}}{\sqrt{u}} du = \frac{e^{2\pi i f(c)}}{\sqrt{2f''(c)}}\int_0^\infty \frac{e^{2\pi i u} du}{\sqrt{u}} - \frac{e^{2\pi i f(c)}}{\sqrt{2f''(c)}}\int_\lambda^\infty \frac{e^{2\pi i u} du}{\sqrt{u}}.$$

We use two methods to estimate the last integral. First,

$$\left|\int_\lambda^\infty \frac{e^{2\pi i u} du}{\sqrt{u}}\right| \le \int_\lambda^{\lambda+1} \frac{du}{\sqrt{u}} + \left|\int_{\lambda+1}^\infty \frac{e^{2\pi i u}}{\sqrt{u}}\right| \ll 1;$$

Moreover, for $\lambda > 0$

$$\left|\int_\lambda^\infty \frac{e^{2\pi i u} du}{\sqrt{u}}\right| \ll \frac{1}{\sqrt{\lambda}} = \frac{1}{\sqrt{f(b)-f(c)}}.$$

Since

$$f(b) - f(c) = \tfrac{1}{2}f''(\xi)(b-c)^2 \gg (b-c)^2 A^{-1},$$
$$|f'(b)| = |f'(b-c+c)| = f''(\xi_1)(b-c) \ll (b-c)A^{-1},$$

it follows that

$$\frac{1}{\sqrt{f(b)-f(c)}} \ll \frac{\sqrt{A}}{b-c} \ll \frac{1}{|f'(b)|\sqrt{A}}.$$

In the same way, we obtain

$$J_1 = \frac{e^{2\pi i f(c)}}{\sqrt{2f''(c)}} \int\limits_0^\infty \frac{e^{2\pi i u} du}{\sqrt{u}} + O\left(\min\left(\frac{1}{|f'(b)|}, \sqrt{A}\right)\right);$$

$$\int\limits_c^b = \frac{\varphi(c)e^{2\pi i f(c)}}{\sqrt{2f''(c)}} \int\limits_0^\infty \frac{e^{2\pi i u} du}{\sqrt{u}} + O(H) + O\left(H \min\left(\frac{1}{|f'(b)|}, \sqrt{A}\right)\right).$$

The first integral in (8) is computed analogously:

$$\int\limits_a^c = \frac{\varphi(c)e^{2\pi i f(c)}}{\sqrt{2f''(c)}} \int\limits_0^\infty \frac{e^{2\pi i u} du}{\sqrt{u}} + O(H) + O\left(H \min\left(\frac{1}{|f'(a)|}, \sqrt{A}\right)\right).$$

Moreover,

$$\int\limits_0^\infty \frac{\cos 2\pi u}{\sqrt{u}} du = \int\limits_0^\infty \frac{\sin 2\pi u}{\sqrt{u}} du = \frac{1}{2}.$$

This completes the proof of the Lemma.   □

Theorem 4. *Let $f(x)$ and $\varphi(x)$ be real functions satisfying the conditions 1 and 2 of Lemma 2. Define the numbers x_n by the equation $f'(x_n) = n$. Then*

$$\sum_{a < x \le b} \varphi(x)e^{2\pi i f(x)} = \frac{1+i}{\sqrt{2}} \sum_{f'(a) \le n \le f'(b)} \frac{\varphi(x_n)}{\sqrt{f''(x_n)}} e^{2\pi i (f(x_n) - n x_n)}$$

$$+ O(H(f'(b) - f'(a)) + O(H\sqrt{A}). \qquad (9)$$

Proof. We obtain (5) from Lemma 1, and we apply Lemma 2 to each integral I_n,

$$I_n = \int\limits_a^b \varphi(x)e^{2\pi i (f(x) - n x)} dx, \quad f'(a) + 1 \le n \le f'(b) - 1,$$

For $f'(a) - 1 \le n < f'(a) + 1$ and $f'(b) - 1 < n \le f'(b) + 1$, we can use the estimate $I_n \ll H\sqrt{A}$. Adding the asymptotic formulas for I_n, we complete the proof of the Theorem.   □

Corollary. *Under the conditions of Theorem 4, we have*

$$\sum_{a < x < h} \varphi(x)e^{2\pi i f(x)} \ll H\left(\frac{b-a}{\sqrt{A}} + \sqrt{A}\right).$$

Proof. The trivial estimate of the right side of (9) gives

$$\sum_{a < x \le b} \varphi(x)e^{2\pi i f(x)} \ll H\left((f'(b) - f'(a) + 1)\sqrt{A}\right) \ll H\left(\frac{b-a}{\sqrt{A}} + \sqrt{A}\right). \quad □$$

It follows easily from this Corollary and from Lemma 2 that

$$\Delta(R) = O(R^{1/3}), \quad \Delta_1(R) = O(R^{1/3} \ln R).$$

However, in the case of the circle problem and the divisor problem, it is possible to estimate nontrivially the new trigonometric sum that appears on the right side of (9), and this allows us to make a more precise statement about the size of $\Delta(R)$ and $\Delta_1(R)$.

Lemma 3. *Let $f(x)$ be a real function on $[a, b]$, and let $q \leq b - a$, where q is a natural number. Then*

$$\left| \sum_{a < n \leq b} e^{2\pi i f(n)} \right| \ll \frac{b-a}{\sqrt{q}} + \sqrt{\frac{b-a}{q}} \sqrt{\sum_{r=1}^{q-1} \left| \sum_{a < n \leq b-r} e^{2\pi i g(n)} \right|},$$

where $g(n) = f(n + r) - f(n)$, and the constant in the $\ll$ sign is absolute.

Proof. For convenience, we set $e^{2\pi i f(n)} = 0$ for $n \leq a$ and $b < n$. Then for any integer m we have

$$S = \sum_{a < n \leq b} e^{2\pi i f(n)} = \sum_n e^{2\pi i f(n+m)},$$

Consequently,

$$S = \frac{1}{q} \sum_n \sum_{m=1}^{q} e^{2\pi i f(n+m)}.$$

Note that the inner sum on the right of the last equation is equal to zero for $n \leq a - q$ and $b < n$, i.e. we can assume that $a - q < n < b$. Applying Cauchy's inequality, we obtain

$$|S|^2 \leq \frac{1}{q^2} \sum_n 1 \sum_n \left| \sum_{m=1}^{q} e^{2\pi i f(n+m)} \right|^2 \leq \frac{1}{q^2} (b - a + q) \sum_n \left| \sum_{m=1}^{q} e^{2\pi i f(n+m)} \right|^2$$

$$\leq \frac{2(b-a)}{q^2} \left\{ 2(b-a)q + 2 \left| \sum_n \sum_{m>s} e^{2\pi i (f(n+m) - f(n+s))} \right| \right\}.$$

Let us consider the multiple sum. We can represent the function in the exponent in the form of a difference $f(v + r) - f(v)$, where $r = 1, 2, \ldots, q - 1$ and $v = [a] + 1, \ldots, [b] - r$, since for the other values of the parameters we obtain zero. For fixed values r and v, the system of equations

$$\begin{cases} n + m = v + r \\ n + s = v, \end{cases} \quad 1 \leq s < m \leq q,$$

has exactly $q - r$ solutions, since $m - s = r$ and the solutions will be the sets $s = 1, m = r + 1, n = v - 1; s = 2, m = r + 2, n = v - 2, \ldots; s = q - r, m = q, n = v - q + r$. In the same way we find

$$\left| \sum_n \sum_{m>s} e^{2\pi i (f(n+m) - f(n+s))} \right| = \left| \sum_v \sum_{r=1}^{q-1} (q - r) e^{2\pi i (f(v+r) - f(v))} \right|$$

$$= \left| \sum_{r=1}^{q-1} (q - r) \sum_{a < v \leq b-r} e^{2\pi i (f(v+r) - f(v))} \right|$$

$$\leq q \sum_{r=1}^{q-1} \left| \sum_{a < n \leq nb-r} e^{2\pi i g(n)} \right|.$$

The Lemma follows from this. $\square$

Theorem 5. *Let $k \geq 2$ and $K = 2^{k-1}$. Let the function $f(x)$ have k continuous derivatives on $[a, b]$, and also satisfy*

$$0 < \lambda_k \leq f^{(k)}(\xi) \leq h\lambda_k \quad \text{for all } \xi \in [a, b].$$

Then the following estimate holds:

$$\left| \sum_{a < n \leq b} e^{2\pi i f(n)} \right| \ll (b - a)\lambda_h^{\frac{1}{2K-2}} h^{\frac{2}{k}} + (b - a)^{1 - \frac{2}{k}}\lambda_h^{-\frac{1}{2K-2}},$$

where the constant in the sign $\ll$ is absolute.

Proof. We shall assume that

$$(b - a)^{-4(1 - \frac{1}{k})} \ll \lambda_k \ll 1,$$

since otherwise the assertion of the Theorem holds trivially. We shall prove the Theorem by induction on k.

Let $k = 2$. If we set $\varphi(x) = 1$, $H = 1$, and $\Delta = 0.5$ in Lemma 1, then we find that

$$\sum_{a < x \leq b} e^{2\pi i f(x)} = \sum_{\alpha - 0,5 \leq u \leq \beta + 0,5} I(n) + O(\ln(\beta - \alpha + 2)),$$

where

$$I(n) = \int_a^b e^{2\pi i (f(x) - ux)} dx,$$

$\alpha = f^{(1)}(a)$ and $\beta = f^{(1)}(b)$. Next we estimate the integral

$$J = \int_a^b e^{2\pi i F(x)} dx,$$

where we assume only that $F^{(2)}(x)$ is continuous on $[a, b]$ and $F^{(2)}(x) \geq \lambda_2 > 0$ for $a \leq x \leq b$. We shall prove that $J \ll \delta = \lambda_2^{-1/2}$. Let $\xi \in [a, b]$ be a root of the equation $F^{(1)}(x) = 0$. Then $J = J_1 + J_2$, where

$$J_1 = \int_a^\xi, \quad J_2 = \int_\xi^b.$$

The integrals J_1 and J_2 can be estimated in the same way. We shall estimate J_1 under the assumption that $\xi - a \geq 2\delta$. We have $J_1 = J_{11} + J_{12}$, where

$$J_{11} = \int_a^{\xi - \delta}, \quad J_{21} = \int_{\xi - \delta}^\xi \ll \delta,$$

Integrating by parts, we find

$$J_{11} = \frac{1}{2\pi i F^{(1)}(x)} e^{2\pi i F(x)} \Big|_a^{\xi - \delta} + \frac{1}{2\pi i} \int_a^{\xi - \delta} \frac{F^{(2)}(x)}{(F^{(1)}(x))^2} e^{2\pi i F(x)} dx.$$

In the last equation, since $F^{(1)}(x)$ is a monotonically increasing function and

$F^{(1)}(\xi) = 0$, we obtain the estimate

$$J_{11} \ll \frac{1}{|F^{(1)}(\xi - \delta)|}.$$

Moreover,

$$|F^{(1)}(\xi - \delta)| = F^{(2)}(\xi_1)\delta \gg \delta^{-1}.$$

Consequently,

$$J_{11} \ll \delta; \quad J_1 \ll \delta; \quad J \ll \delta.$$

We can treat the case $\xi \notin [a, b]$ and $|F^{(1)}| > 0$ in the same way.
Now we apply the estimate $I(n) \ll \delta$ to each integral $I(n)$ and find that

$$\sum_{a < x \leq b} e^{2\pi i f(x)} \ll (\beta - \alpha + 1)\delta + \ln(\beta - \alpha + 2) \ll (f^{(1)}(b) - f^{(1)}(a))\lambda_2^{-\frac{1}{2}} + \lambda_2^{-\frac{1}{2}}$$

$$\ll (b - a)h\lambda^{\frac{1}{2}} + \lambda_2^{-\frac{1}{2}}.$$

This proves the Theorem in the case $k = 2$.
Now let $k > 2$, and suppose that the Theorem holds for $k - 1$. We shall prove
the Theorem for k. Let $q = [\lambda_k^{-1/(k-1)}]$ and apply

Lemma 3. *Since*

$$\lambda_k \gg (b - a)^{-4(1 - k)}, \quad K > 4,$$

it follows that

$$1 \leq q \leq \lambda_k^{-\frac{1}{k-1}} < b - a,$$

and the Lemma gives

$$\left| \sum_{a < n \leq} e^{2\pi i f(n)} \right| \leq c_0 (b - c)\lambda_k^{\frac{1}{2^{K-2}}}$$

$$+ c_0 \left((b - a)\lambda_k^{\frac{1}{k-1}} \sum_{r=1}^{q-1} \left| \sum_{a < n \leq b - r} e^{2\pi i g(n)} \right| \right)^{\frac{1}{2}}$$

where $g(n) = f(n + r) - f(n)$. Since

$$g^{(k-1)}(x) = f^{(k-1)}(x + r) - f^{(k-1)}(x) = rf^{(k)}(\xi)$$

for some $\xi \in [a, b]$, then

$$r\lambda_k \leq g^{(k-1)}(x) \leq rh\lambda_k, \quad x \in [a, b].$$

Therefore, the estimate in the induction hypothesis can be applied to the inner
sum: $K_1 = 2^{k-2}$, $2K_1 = K$, and

$$\sigma_r = \left| \sum_{a < n \leq b - r} e^{2\pi i g(n)} \right| \leq c_{k-1}(b - a)(r\lambda_k)^{\frac{1}{2K_1 - 2}} h^{\frac{2}{K_1}} + c_{k-1}(b - a)^{1 - \frac{1}{K_1}}(r\lambda_k)^{-\frac{1}{2K_1 - 2}}.$$

From this we find

$$\sum_{r-1}^{q-1} \sigma_r \le c_{k-1}(b-a)q^{1+\frac{1}{K-2}}\lambda_k^{\frac{1}{K-2}}h^{\frac{4}{k}} + 2c_{k-1}(b-a)^{1-\frac{4}{k}}q^{1-\frac{1}{K-2}}\lambda_k^{-\frac{1}{K-2}}$$

$$\le c_{k-1}(b-a)h^{\frac{4}{k}} + 2c_{k-1}(b-a)^{1-\frac{4}{k}}\lambda_k^{-\frac{2}{K-1}}.$$

In the same way we obtain an estimate for our sum

$$\left| \sum_{a<n\le b} e^{2\pi i f(n)} \right| \le c_0(b-a)\lambda_k^{\frac{1}{2K-2}} + c_0 c_{k-1}^{\frac{1}{2}}(b-a)\lambda_k^{\frac{1}{2K-2}}h^{\frac{2}{k}}$$

$$+ c_0(2c_{k-1})^{\frac{1}{2}}(b-a)^{1-\frac{2}{k}}\lambda^{-\frac{1}{2K-2}}$$

$$\le c_k\left((b-a)\lambda_k^{\frac{1}{2K-2}}h^{\frac{2}{k}} + (b-a)^{1-\frac{2}{k}}\lambda_k^{-\frac{1}{2K-2}} \right),$$

where $c_k \le \max(c_0 + c_0 c_{k-1}^{1/2}, c_0(2c_{k-1})^{1/2})$.

Without loss of generality, we can assume $c_1 = 4c_0^2$. Then $c_k \le c_1$, and the Theorem is proved.   □

§4. Integer Points in a Circle and Under a Hyperbola

We shall apply the previous theorem to the problem of integer points.

Theorem 6. *The following asymptotic formula holds for* $K(R)$:

$$K(R) = \pi R + O(R^{\frac{1}{3}-\frac{1}{246}}\ln R).$$

Proof. We compute σ, where

$$\sigma = \sum_{0<x\le \sqrt{R/2}} \{\sqrt{R-x^2}\}.$$

Let $r = [\ln R] \ge 1$, $\Delta = R^{-7/41}$. Let α and β be arbitrary real numbers satisfying $0 \le \alpha < \beta < 1$, and let $\psi(x)$ be the function of Lemma A, corresponding to the given r, Δ, α, and β. By Lemma B, it is sufficient to estimate σ_0,

$$\sigma_0 = \sum_{\substack{m=-\infty \\ m\ne 0}}^{+\infty} g(m)U_m,$$

where

$$U_m = \sum_{0<x<\sqrt{R/2}} c^{2\pi i m\sqrt{R-x^2}}, \quad |g(m)| \le \min\left(\frac{1}{\pi|m|}, \frac{1}{\pi|m|}\left(\frac{r}{\pi|m|\Delta}\right)^r \right).$$

We have

$$\left| \sum_{|m|>\Delta^{-1}\ln R} g(m)U_m \right| \le 2\sqrt{\frac{R}{2}} \sum_{m>\Delta^{-1}\ln R} \frac{1}{\pi m}\left(\frac{r}{\pi m\Delta}\right)^r \ll 1.$$

It remains to estimate the sum over m for $1 \le |m| \le \varDelta^{-1} \ln R$. To the trigono-metric sum

$$\bar{U}_m = \sum_{0 < x \le \sqrt{R/2}} e^{-2\pi i m \sqrt{R - x^2}},$$

We apply Theorem 4 with $H = 1$, $f(x) = -m\sqrt{R - x^2}$, $U = \sqrt{R/2}$, $a = 0$, $b = \sqrt{R/2}$, $A = \sqrt{R}/m$. All of the conditions of the Theorem are satisfied, and we have

$$f'(x) = \frac{mx}{\sqrt{R - x^2}}; \quad x_n = \frac{n\sqrt{R}}{\sqrt{n^2 + m^2}};$$

$$g(n) = f(x_n) - n(x_n) = -\sqrt{R(n^2 + m^2)};$$

$$f''(x) = \frac{mR}{(R - x^2)^{3/2}}; \quad \sqrt{f''(x_n)} = \frac{(n^2 + m^2)^{3/4}}{m^4 \sqrt{R}};$$

$$f'(0) = 0; \quad f'\left(\sqrt{\frac{R}{2}}\right) = m;$$

$$\bar{U}_m = \sum_{0 < n \le m} \frac{1 + i}{\sqrt{2}} \frac{m^4 \sqrt{R}}{(m^2 + n^2)^{3/4}} e^{-2\pi i \sqrt{R(n^2 + m^2)}} + O(\sqrt[4]{R}/\sqrt{m}).$$

From this and from the estimate $|g(m)| \le 1/\pi|m|$, we find

$$\left| \sum_{0 < |m| \le \varDelta^{-1} \ln R} g(m) U_m \right| \ll \sqrt[4]{R} \sum_{10 < m \le \varDelta^{-1} \ln R} \left| \sum_{0 < n \le m} (m^2 + n^2)^{-\frac{3}{4}} e^{2\pi i \sqrt{R(n^2 + m^2)}} \right|$$

$$+ \sqrt[4]{R}. \tag{10}$$

Let us estimate the inner sum. We apply Lemma C (partial summation) with

$$c_n = e^{2\pi i \sqrt{R(n^2 + m^2)}}, \quad f(n) = (m^2 + n^2)^{-\frac{3}{4}}.$$

Then

$$S = \sum_{10 < n \le m} (m^2 + n^2)^{-\frac{3}{4}} e^{2\pi i \sqrt{R(n^2 + m^2)}}$$

$$= \frac{3}{2} \int_{10}^{m} C(x) x (m^2 + x^2)^{-\frac{7}{4}} dx + C(m)(2m^2)^{-\frac{3}{4}},$$

where $C(x) = \sum_{10 < n \le x} e^{2\pi i \sqrt{R(n^2 + m^2)}}, \quad x \le m$.

To estimate the sum $C(x)$, we apply Theorem 5, after first dividing the interval of summation $(10, x]$ into $\ll \ln R$ intervals of the form $(N, N_1]$, where $N_1 \le 2N$. First we find $f^{(v)}(x)$:

$$f^{(v)}(n) = -60\sqrt{R} m^2 n (n^2 + m^2)^{-\frac{9}{2}} \left(n^2 - \frac{3}{4} m^2\right).$$

It is possible that the zero $n_0 = (\sqrt{3}/2)m$ of $f^{(v)}(x)$ lies inside the interval of summation $(N, N_1]$. To consider the most general case, we divide the interval $(N, N_1]$ into $\ll \ln R$ intervals $E_0, E_1, E_2, \ldots, E_{-1}, E_{-2}, \ldots$, where

$$E_0 = \{n: n_0 - 1 \leq n \leq n_0 + 1\},$$

$$E_v = \{n: n_0 + 2^{v-1} < n \leq \min(n_0 + 2_0^v N_1)\}, \quad v = 1, 2, \ldots;$$

$$E_{-v} = \{n: \min(N, n_0 - 2^v) \leq n < n_0 - 2^{v-1}\}, \quad v = 1, 2, \ldots.$$

The length of E_0 is 2. The length of $E_{\pm v}$ does not exceed 2^{v-1}, where $1 \leq v \leq \log_2 N + 1$. Moreover, for $n \in E_{\pm v}, v \geq 1$,

$$|f^{(v)}(n)| \asymp \sqrt{R} m^{-6} N_{2^v} = \lambda_5.$$

Therefore, setting $k = 5$ in Theorem 5, we find for $v \geq 1$:

$$\sum_{n \in E_{\pm v}} e^{2\pi i \sqrt{R(n^2 + m^2)}} \ll 2^v (\sqrt{R} m^{-6} N_2^v)^{\frac{1}{30}} + 2^{v(1-\frac{1}{8})} (\sqrt{R} m^{-6} N_{2^v})^{-\frac{1}{30}};$$

$$\mathbb{D}(x) \ll m^{\frac{13}{15}} R^{\frac{1}{60}}; \quad S \ll m^{-\frac{19}{30}} R^{\frac{1}{60}}.$$

Inserting this estimate in (10), we have

$$\left| \sum_{0 < |m| \leq \Delta^{-1} \ln R} g(m) U_m \right| \ll R^{\frac{1}{3} - \frac{1}{246}} \ln R.$$

The Theorem follows from this and Lemma B.   □

Theorem 7. *The number $L(R)$ of integer points with positive coordinates under the hyperbola $xy = R$ is given by the asymptotic formula*

$$L(R) = R(\ln R + 2\gamma - 1) + O(R^{\frac{1}{3} - \frac{1}{246}} \ln^2 R).$$

Proof. It suffices to prove the formula

$$\sum_{x \leq \sqrt{R}} \left\{ \frac{R}{x} \right\} = \frac{1}{2} \sqrt{R} + O(R^{\frac{1}{3} - \frac{1}{246}} \ln^2 R).$$

The points $2^{-k}\sqrt{R}$, where $k = 0, 1, 2, \ldots$, divide the interval

$$(R^{\frac{1}{3} - \frac{1}{246}}, \sqrt{R})$$

into $\ll \ln R$ intervals of the form $(a, 2a)$, where $2a \leq \sqrt{R}$ and $a > R^{1/3 - 1/246}$. We shall prove the inequality

$$\left| \sum_{a < x \leq 2a} \left\{ \frac{R}{x} \right\} - \frac{a}{2} \right| \ll R^{\frac{1}{3} - \frac{1}{246}} \ln R.$$

We use trigonometric sums, and apply Lemma A with $r = [\ln R]$. As in Theorem 6, it will be necessary to estimate the sum

$$\sum_{0 < m \leq \Delta^{-1} \ln R} \frac{1}{m} |U(m)|,$$

where $\varDelta$ is a parameter, and

$$U(m) = \sum_{a < x \le 2a} e^{-2\pi i \frac{mR}{x}}.$$

We shall estimate the sum $U(m)$ in various ways, depending on the value of a, or, in other words, depending on the length of the interval of summation of the numbers $\{R/x\}$.

1. Let

$$R^{\frac{1}{3} - \frac{1}{246}} < a \le R^{\frac{7}{17}}.$$

We take

$$\varDelta = R^{\frac{1}{3}} a^{-\frac{4}{3}} \ln^{\frac{2}{3}} R < \frac{1}{10},$$

and estimate $U(m)$ by Theorem 5, in which we set $k = 3$, $K = 4$, $\lambda_3 = mR/a^4$, and $h \ll 1$. We find that

$$U(m) \ll a(mR/a^4)^{\frac{1}{6}} + a^{1 - \frac{1}{2}}(mR/a^4)^{-\frac{1}{6}}; \qquad \sum_{0 < m \le \varDelta^{-1} \ln R} \frac{1}{m} |U(m)| \ll R^{\frac{1}{3} - \frac{1}{246}} \ln R$$

2. Now let $R^{7/17} \le a \le \sqrt{R}$. In this case we use Theorem 4 to pass from $U(m)$ to a shorter sum, and then apply Theorem 5. For R sufficiently large, we set

$$\varDelta = R^{\frac{11}{41}} a^{-\frac{36}{41}} \ln^2 R < \frac{1}{10}$$

(We assume that T is sufficiently large.) Setting $f(x) = -mR/x$ in Theorem 4, we find

$$f'(x) = mR/x^2, \quad f''(x) = -2mR/x^3; \quad A = \frac{a^3}{mR} \gg \frac{a^3}{R} \varDelta \ln^{-1} R \gg 1;$$

$$x_n = \sqrt{mR/n}, \quad g(n) = -2\sqrt{mnR}, \quad \sqrt{|f''(x_n)|} = \sqrt{2} n^{\frac{3}{4}} (mR)^{-\frac{1}{4}};$$

$$f'(a) = mR/a^2, \quad f'(2a) = mR/4a^2;$$

$$U(m) = \frac{1 + i}{2} (mR)^{\frac{1}{4}} \sum_{\frac{mR}{4a^2} \le n \le \frac{mR}{a^2}} n^{-\frac{3}{4}} e^{-2\pi i 2 \sqrt{mnR}} + O\left(\sqrt{\frac{a^3}{mR}}\right).$$

The contribution of the remainder term in the last formula has order of magnitude

$$\ll \sum_{m \le \varDelta^{-1} \ln R} \frac{1}{m} \sqrt{\frac{a^3}{mR}} \ll R^{-\frac{1}{2}} a^{\frac{3}{2}} < R^{\frac{1}{4}}.$$

Further, applying partial summation, we get

$$\sum_{\frac{mR}{4a^2} \le n \le \frac{mR}{a^2}} n^{-\frac{3}{4}} e^{-2\pi i 2 \sqrt{mnR}} = \frac{3}{4} \int_{mR/4a^2}^{mR/a^2} x^{-\frac{7}{4}} C(x) dx + C\left(\frac{mR}{a^2}\right)\left(\frac{mR}{a^2}\right)^{-\frac{3}{4}},$$

where

$$C(x) = \sum_{\frac{mR}{4a^2} \le n \le x} e^{-2\pi i 2\sqrt{mnR}}.$$

We estimate $C(x)$ by Theorem 5, with $f(n) = 2\sqrt{mnR}$ and $k = 5$. Then $K = 16$, and

$$\lambda_5 = (mR)^{\frac{1}{2}}\left(\frac{mR}{a^2}\right)^{-\frac{9}{2}} = (mR)^{-4}a^9;$$

$$C(x) \ll (mR)^{\frac{13}{15}}a^{-\frac{17}{10}} + (mR)^{\frac{121}{120}}a^{-\frac{41}{20}};$$

$$U(m) \ll (mR)^{\frac{11}{30}}a^{-\frac{1}{3}} + (mR)^{\frac{61}{120}}a^{-\frac{11}{20}};$$

$$\sum_{m \le \varDelta^{-1}\ln R} \frac{1}{m}|U(m)| \ll R^{\frac{1}{3}-\frac{1}{246}}\ln R.$$

The Theorem follows from these estimates. □

Exercises

1. Let the function $f''(x)$ be continuous on the interval $a \le x \le b$ and satisfy the conditions

$$f''(x) \ge 1/A, \quad |f'(x)| \le D.$$

for some $A \ge 1$ and $D \ge 1$. Then

$$\sum_{a < x \le b} e^{2\pi i f(x)} \ll (f'(b) - f'(a))\sqrt{A} + \sqrt{A} + \ln((b-a)D).$$

2. Let $b - a \ll A$, $A \gg 1$, and let the function $f''(x)$ be continuous on $[a, b]$. Moreover, let

$$f''(x) \gg 1/A, \quad 0 < f'(x) \ll 1, \quad a \le x \le b.$$

Then the following asymptotic equations hold:

$\alpha)\ \displaystyle\sum_{a < x \le b} \{f(x)\} = \frac{b-a}{2} + O(A^{\frac{2}{3}}),$

$\beta)\ T_f(a, b) = \displaystyle\int_a^b f(x)dx + \rho(b)f(b) - \rho(a)f(a) - \frac{b-a}{2} + O(A^{\frac{2}{3}}),$

where $\rho(x) = 1/2 - \{x\}$ and $T_f(a, b)$ is the number of integer points in the curvilinear trapezoid $a < x \le b, 0 < y \le f(x)$.

3. Construct a curve $y = f(x)$ such that the conditions of exercise 2 are satisfied, and such that on the curve there are $\gg A^{2/3}$ integer points. It can be

proved in the same way that the remainder term in formula β) cannot be replaced by $o(a^{2/3})$.

4. Let $V(R)$ be the number of integer points in the sphere $X^2 + Y^2 + Z^2 \leq R^2$. Then the following asymptotic formula holds:

$$V(R) = \frac{4}{3}\pi R^3 + O(R^{\frac{3}{2}}\ln^2 R).$$

5. Prove that the following inequalities hold for $a \leq \sqrt{|t|}$:

$\alpha)\ \sum_{a < n \leq 2a} n^{it} \ll \sqrt{a}|t|^{\frac{1}{6}},$

$\beta)\ \sum_{a < n \leq 2a} n^{it} \ll \sqrt{a}|t|^{\frac{1}{6} - \frac{1}{252}}.$

Chapter II. Entire Functions of Finite Order

This chapter provides background information from the theory of entire functions that will be used later in the book.

§1. Infinite Products. Weierstrass's Formula

We introduce the concept of infinite products.

Definition 1. Let $u_1, u_2, \ldots, u_n, \ldots$ be an infinite sequence of complex numbers different from -1. An *infinite product* is an expression of the form

$$\prod_{n=1}^{\infty} (1 + u_n) = (1 + u_1)(1 + u_2) \ldots (1 + u_n) \ldots \tag{1}$$

An expression of the form

$$\prod_{n=1}^{k} (1 + u_n) = (1 + u_1) \ldots (1 + u_k) = v_k \tag{2}$$

is called a *partial product*.

Definition 2. If the sequence (2) of numbers v_k converges as $k \to \infty$ to a number $v \neq 0$, then we say that the infinite product (1) *converges* and is equal to v, i.e.

$$v = \lim_{k \to \infty} v_k = \prod_{n=1}^{\infty} (1 + u_n). \tag{3}$$

If the sequence v_k does not converge or if $v = 0$, then the infinite produce (1) is called *divergent*.

For most applications, the following test for convergence is sufficient.

Theorem 1. *If the series*

$$u_1 + u_2 + \ldots + u_n + \ldots \tag{4}$$

converges absolutely, then the product (1) also converges.

Proof. Since the series

$$|u_1| + |u_2| + \ldots + |u_n| + \ldots,$$

converges, it follows that $\lim_{n \to \infty} |u_n| = 0$. Consequently, without loss of generality, we can assume that $|u_n| \leq 1/2$ for $n = 1, 2, \dots$. Let us suppose first that $u_n = \alpha_n$ is a real number for all $n = 1, 2, \dots$. Then $|\ln(1 + u_n)| \leq 2|u_n|$. From this follows the convergence of the sequence $\ln(1 + u_1) + \dots + \ln(1 + u_n) = \ln(1 + u_1) \dots (1 + u_n)$, and, consequently, of the product (1).

Now let u_n be arbitrary complex numbers. We must prove that as $n \to \infty$ the following two sequences converge:

$$|v_n| = |(1 + u_1) \dots (1 + u_n)| = |1 + u_1| \dots |1 + u_n|, \tag{5}$$

$$\arg v_n = \arg(1 + u_1) \dots (1 + u_n) = \arg(1 + u_1) + \dots + \arg(1 + u_n). \tag{6}$$

The convergence of the sequence (5) is equivalent to the convergence of the sequence $|v_n|^2$. But

$$|1 + u_n|^2 = |1 + \alpha_n + i\beta_n|^2 = 1 + \alpha_n^2 + \beta_n^2 + 2\alpha_n,$$

$$\alpha_n = \operatorname{Re} u_n, \quad \beta_n = \operatorname{Im} u_n,$$

and since $|\alpha_n^2 + \beta_n^2 + 2\alpha_n| \leq |u_n|^2 + 2|u_n|$, the convergence of $|v_n|^2$ follows from what has just been proved. The convergence of (6) follows from the fact that for sufficiently large n_0 and $n > n_0$

$$|\arg(1 + u_n)| = \left| \arc\sin \frac{\beta_n}{\sqrt{(1 + \alpha_n)^2 + \beta_n^2}} \right| < \pi |\beta_n|.$$

This proves Theorem 1.   □

We proceed to the study of infinite products of functions analytic in a domain.

Theorem 2. *Let $u_n(s)$ be an infinite sequence of functions analytic in a domain G. We assume that*

a) $u_n(s) \neq -1$ *for $n = 1, 2, \dots$ and $s \in G$,*
b) $|u_n(s)| \leq a_n$ *for $n = 1, 2, \dots$ and $s \in G$,*
c) *the series of numbers $\sum_{n=1}^{\infty} a_n$ converges.*

Then the product

$$\prod_{n=1}^{\infty} (1 + u_n(s)) \tag{7}$$

converges for all $s \in G$, and the function $v(s)$ defined by the equation

$$v(s) = \prod_{n=1}^{\infty} (1 + u_n(s)),$$

is analytic in the domain G. Moreover, $v(s) \neq 0$ for $s \in G$.

Proof. The convergence of (7) for $s \in G$ follows from Theorem 1. To prove that $v(s)$ is analytic in G, it suffices to show that the sequence of analytic functions $v_k(s)$ converges uniformly to $v(s)$ for $s \in G$, where $v_k(s)$ is defined by

$$v_k(s) = \prod_{n=1}^{k} (1 + u_n(s)).$$

We set $\quad \prod_{n=1}^{\infty} (1 + a_n) = p, \quad (1 + a_1)\ldots(1 + a_n) = p_n.$

We shall first show that for any $s \in G$

$$\left| \frac{v(s)}{v_n(s)} - 1 \right| \leq \frac{p}{p_n} - 1. \tag{8}$$

Clearly, if $k \geq 1$, then

$$\left| \frac{v_{n+k}(s)}{v_n(s)} - 1 \right| = |(1 + u_{n+1}(s))\ldots(1 + u_{n+k}(s)) - 1|$$

$$= |u_{n+1}(s) + \ldots + u_{n+k}(s) + u_{n+1}(s)u_{n+2}(s) + \ldots$$

$$+ u_{n+1}(s)\ldots u_{n+k}(s)| \leq a_{n+1} + \ldots + a_{n+k}$$

$$+ a_{n+1}a_{n+2} + \ldots + a_{n+1}\ldots a_{n+k} = \frac{p_{n+k}}{p_n} - 1.$$

We obtain (8) by passing to the limit as $k \to \infty$. Then we have

$$|v(s) - v_n(s)| = |v_n(s)| \left| \frac{v(s)}{v_n(s)} - 1 \right| \leq p_n \left(\frac{p}{p_n} - 1 \right) = p - p_n < \varepsilon$$

for $n \geq n_0(\varepsilon)$ and all $s \in G$. This proves the Theorem. $\qquad \square$

Definition 3. A function $f(s)$ that is analytic in every finite part of the complex s-plane is called *entire*.

Next, we shall obtain two theorems on entire functions. The first theorem proves the existence of an entire function having as its zeros only the numbers in a given infinite sequence. The second theorem (a generalization of the Fundamental Theorem of Algebra) gives the decomposition of an entire function as an infinite product over its zeros.

Theorem 3. *Let* $a_1, \ldots, a_n, \ldots$ *be an infinite sequence of complex numbers such that*

$$0 < |a_1| \leq |a_2| \leq \ldots \leq |a_n| \leq \ldots$$

and

$$\lim_{n \to \infty} \frac{1}{|a_n|} = 0.$$

Then there exists an entire function $G(s)$ *that has as its zeros only the numbers* a_n. *If some of the* a_n *are equal, then the zero* a_n *of* $G(s)$ *has the appropriate multiplicity.*

Proof. For $n = 1, 2, \ldots$, we set

$$u_n = u_n(s) = \left(1 - \frac{s}{a_n} \right) e^{\frac{s}{a_n} + \frac{1}{2}\left(-\frac{s}{a_n}\right)^2 + \ldots + \frac{1}{n-1}\left(\frac{s}{a_n}\right)^{n-1}}$$

and we consider the infinite product

$$\prod_{n=1}^{\infty} u_n(s). \tag{9}$$

We shall prove that this product converges at every complex number $s \neq a_n$, $n = 1, 2, \ldots$, and is an entire function $G(s)$ with zeros $a_1, a_2, \ldots, a_n, \ldots$. For this, we consider the circle C of radius $|a_n|$ and the infinite product $\prod_{r=n}^{\infty} u_r(s)$. We shall show that the latter product converges in C to an analytic function. Then (9) will also be an analytic function, which, inside the circle C, has only the zeros a_i with $|a_i| < |a_n|$. Since $|a_n| \to \infty$, the Theorem will be proved. For $|s| < |a_n|$ and $r \geq n$,

$$\ln u_r(s) = \ln\left(1 - \frac{s}{a_r}\right) + \frac{s}{a_r} + \frac{1}{2}\left(\frac{s}{a_r}\right)^2 + \ldots + \frac{1}{r-1}\left(\frac{s}{a_r}\right)^{r-1}$$

$$= -\frac{1}{r}\left(\frac{s}{a_r}\right)^r - \frac{1}{r+1}\left(\frac{s}{a_r}\right)^{r+1} - \ldots$$

and so

$$u_r(s) = e^{-\frac{1}{r}\left(\frac{s}{a_r}\right)^r - \frac{1}{r+1}\left(\frac{s}{a_r}\right)^{r+1} - \ldots}$$

Thus, it suffices to show the absolute convergence of the series

$$\sum_{r=n}^{\infty}\left[\frac{1}{r}\left(\frac{s}{a_r}\right)^r + \frac{1}{r+1}\left(\frac{s}{a_r}\right)^{r+1} + \ldots\right] \tag{10}$$

for $|s| < |a_n|$. But for any $0 < \varepsilon < 1/2$ and $|s| \leq (1 - \varepsilon)|a_n|$, we have

$$\left|\frac{1}{r}\left(\frac{s}{a_r}\right)^r + \frac{1}{r+1}\left(\frac{s}{a_r}\right)^{r+1} + \ldots\right| \leq \frac{1}{r}(1 - \varepsilon)^r$$

$$+ \frac{1}{r+1}(1 - \varepsilon)^{r+1} + \ldots < \frac{(1 - \varepsilon)^r}{\varepsilon r}.$$

From this follows the absolute convergence of (10) in the domain $|s| \leq (1 - \varepsilon)|a_n|$, and so the function (9) is analytic in the circle C. This proves the Theorem. $\square$

Corollary 1 (Weierstrass's formula). *Let $a_1, a_2, \ldots, a_n, \ldots$ be a sequence of complex numbers satisfying the conditions of Theorem 3. Then the function $G(s)$*

$$G(s) = s^m \prod_{n=1}^{\infty}\left(1 - \frac{s}{a_n}\right)e^{\frac{s}{a_n} + \frac{1}{2}\left(\frac{s}{a_n}\right)^2 + \ldots + \frac{1}{n-1}\left(\frac{s}{a_n}\right)^{n-1}}$$

is entire and its only zeros are the numbers $0, a_1, a_2, \ldots, a_n, \ldots$.

Corollary 2. *Let the sequence of numbers* $a_1, a_2, \ldots, a_n, \ldots$ *satisfy the conditions of Theorem 3, and suppose that there exists an integer* $p \geq 0$ *such that the series*

$$\sum_{n=1}^{\infty} \frac{1}{|a_n|^{p+1}}.$$

converges. Then the function $G_1(s)$ *defined by*

$$G_1(s) = \prod_{n=1}^{\infty} \left(1 - \frac{s}{a_n}\right) e^{\frac{s}{a_n} + \frac{1}{2}\left(\frac{s}{a_n}\right)^2 + \ldots + \frac{1}{p}\left(\frac{s}{a_n}\right)^p},$$

satisfies Theorem 3.

Proof. For $|s| \leq (1 - \varepsilon)|a_n|$, the series

$$\sum_{r=n}^{\infty} \left[\frac{1}{p+1}\left(\frac{s}{a_r}\right)^{p+1} + \frac{1}{p+2}\left(\frac{s}{a_r}\right)^{p+2} + \ldots\right]$$

is majorized by the series

$$\sum_{r-n}^{\infty} \frac{(1-\varepsilon)^{p+1}}{(p+1)\varepsilon} \cdot \frac{|a_n|^{p+1}}{|a_r|^{p+1}} < +\infty.$$

$\square$

Theorem 4. *Every entire function* $G(s)$ *can be represented in the form*

$$G(s) = e^{H(s)} s^m \prod_{n=1}^{\infty} \left(1 - \frac{s}{a_n}\right) e^{\frac{s}{a_n} + \frac{1}{2}\left(\frac{s}{a_n}\right)^2 + \ldots + \frac{1}{n-1}\left(\frac{s}{a_n}\right)^{n-1}}, \quad (*)$$

where $H(s)$ *is an entire function and the numbers* $0, a_1, a_2, \ldots, a_n, \ldots$ *are the zeros of* $G(s)$, *arranged in order of increasing modulus. If, moreover, the sequence* a_n *for* $n = 1, 2, \ldots$ *satisfies the conditions of Corollary 2, then*

$$G(s) = e^{H(s)} s^m \prod_{n=1}^{\infty} \left(1 - \frac{s}{a_n}\right) e^{\frac{s}{a_n} + \frac{1}{2}\left(\frac{s}{a_n}\right)^2 + \ldots + \frac{1}{p}\left(\frac{s}{a_n}\right)^p}.$$

Proof. Since the zeros of $G(s)$ cannot have a limit point, then they can be arranged in order of increasing modulus. By Theorem 3 we can construct an entire function $G_1(s)$ that has exactly the same zeros as $G(s)$. Setting

$$\varphi(s) = \frac{G(s)}{G_1(s)} \quad \text{for} \quad s \neq a_n, \quad \varphi(a_n) = \lim_{s \to a_n} \varphi(s),$$

we see that $\varphi(s)$ is an entire function nowhere equal to zero, and so the logarithm of $\varphi(s)$ is an entire function. Then $\varphi(s) = e^{H(s)}$, where $H(s)$ is an entire function. This proves the first part of the Theorem, and the second part is proved similarly.

$\square$

§2. Entire Functions of Finite Order

We introduce a series of definitions that will be needed later.

Definition 4. Let $G(s)$ be an entire function, and let
$$M(r) = M_G(r) = \max_{|s| = r} |G(s)|.$$
If there exists $a > 0$ such that
$$M(r) < e^{r^a} \quad \text{for} \quad r > r_0(a) > 0, \tag{11}$$
then $G(s)$ is called an *entire function of finite order*. In this case, $\alpha = \inf a$ is called the order of $G(s)$. If (11) cannot be satisfied for any $a > 0$, then we say that the order of $G(s)$ is ∞.

Definition 5. Let $s_1, s_2, \ldots, s_n, \ldots$ be a sequence of complex numbers such that
$$0 < |s_1| \le |s_2| \le \ldots \le |s_n| \le \ldots \tag{12}$$
If there exists $b > 0$ such that
$$\sum_{n=1}^{\infty} |s_n|^{-b} < +\infty, \tag{13}$$
then we say that the sequence (12) has a *finite exponent of convergence*. In this case $\beta = \inf b$ is called the *exponent of convergence* of (12). If (13) is not satisfied for any $b > 0$, then we say that the exponent of convergence of (12) is equal to ∞.

The fundamental result of this section is the following.

Theorem 5. *Let $G(s)$ be an entire function of finite order α with $G(0) \ne 0$. Let s_n be the sequence of all zeros of $G(s)$, with $0 < |s_1| \le |s_2| \le \ldots \le |s_n| \le \ldots$. Then the sequence s_n has finite exponent of convergence $\beta \le \alpha$, and*
$$G(s) = e^{g(s)} \prod_{n=1}^{\infty} \left(1 - \frac{s}{s_n}\right) e^{\frac{s}{s_n} + \frac{1}{2}\left(\frac{s}{s_n}\right)^2 + \ldots + \frac{1}{p}\left(\frac{s}{s_n}\right)^p},$$
where $p \ge 0$ is the smallest integer such that
$$\sum_{n=1}^{\infty} |s_n|^{-(p+1)} < +\infty.$$
Moreover, $g(s)$ is a polynomial of degree $g \le \alpha$, and $\alpha = \max(g, \beta)$. If for any $c > 0$ there exists an infinite sequence of real numbers $r_1, r_2, \ldots, r_n, \ldots$ such that $r_n \to \infty$ and
$$\max|G(s)| > e^{cr_n^\alpha}, \quad |s| = r_n, \quad n = 1, 2, \ldots,$$
then $\alpha = \beta$ and the series $\sum_{n=1}^{\infty} |s_n|^{-\beta}$ converges.

To prove Theorem 5, we require a series of auxiliary results, which are given in Lemmas 1–5 below. In these Lemmas we shall assume that the hypotheses of Theorem 5 are satisfied, and we shall also use the notation of this Theorem.

Lemma 1. *Let $0 < r < R$ and let m be the number of zeros of $G(s)$ in the circle $|s| < r$. Then*

$$\left(\frac{R}{r}\right)^m \leq \frac{M(R)}{|G(0)|}, \quad \text{where} \quad M(R) = \max_{|s| = R} |G(s)|.$$

Proof. Consider the function $F(s)$ defined by

$$F(s) = G(s) \prod_{n=1}^{m} \frac{R^2 - s\bar{s}_n}{R(s - s_n)}, \quad s \neq s_n,$$

$$F(s_n) = \lim_{s \to s_n} G(s) \prod_{n=1}^{m} \frac{R^2 - s\bar{s}_n}{R(s - s_n)},$$

where $s_1, s_2, \ldots, s_n$ are the zeros of $G(s)$ in the circle $|s| < r$.

The function $F(s)$ is analytic in the circle $|s| \leq R$, and for $|s| = R$

$$|F(s)| = |G(s)|.$$

It follows from the maximum principle that

$$|F(0)| = |G(0)| \prod_{n=1}^{m} \frac{R}{|s_n|} \leq \max_{|s| = R} |F(s)| = M(R).$$

This immediately implies the Lemma. □

Corollary. *Let m be the number of zeros of the function $G(s)$ in the circle of radius $r = R/2$. Then*

$$m \leq \frac{1}{\ln 2} \ln \frac{M(R)}{|G(0)|}.$$

Lemma 2. *Let $N(r)$ be the number of zeros of the function $G(s)$ in the circle of radius r. Then for any $\varepsilon > 0$ there is a constant $C = C(\varepsilon) > 0$ such that*

$$N(r) < Cr^{\alpha + \varepsilon}.$$

Moreover, $\beta \leq \alpha$.

Proof. The first inequality follows from Lemma 1 and the definition of the order of $G(s)$. We shall prove the convergence of the series $\sum_{n=1}^{\infty} |s_n|^{-b}$ for any $b > \alpha$. This implies the second assertion of the Lemma. Since $n < c|s_n|^{\alpha + \varepsilon}$ for any $\varepsilon > 0$, i.e. $|s_n|^{-b} \leq \frac{b}{c^{\alpha + \varepsilon} n^{\frac{b}{\alpha + \varepsilon}}}$, and since $b > \alpha$, then $\frac{b}{\alpha + \varepsilon} > 1$ for sufficiently small $\varepsilon > 0$, and so the series converges. □

Lemma 3. *Let s_n be the sequence (12) with finite exponent of convergence β, and let $p \geq 0$ be the smallest integer such that $\sum\limits_{n=1}^{\infty} |s_n|^{-(p+1)} < +\infty$. Let $P(s)$ be the entire function defined by the equation*

$$P(s) = \prod_{n=1}^{\infty} \left(1 - \frac{s}{s_n}\right) e^{\frac{s}{s_n} + \frac{1}{2}\left(\frac{s}{s_n}\right)^2 + \ldots + \frac{1}{p}\left(\frac{s}{s_n}\right)^p}. \tag{14}$$

The $P(s)$ has order β. Moreover, if $|s_n| \to \infty$ and $\sum\limits_{n=1}^{\infty} |s_n|^{-\beta} < +\infty$, then

$$|P(s)| \leq e^{cr^\beta}, \quad |s| = r.$$

Proof. Let α be the order of $P(s)$. It follows from Lemma 2 that $\beta \leq \alpha$, and so it suffices to prove that $\alpha \leq \beta + \varepsilon$ for every $\varepsilon > 0$, i.e. that $\ln|P(s)| < c(\varepsilon)|s|^{\beta+\varepsilon}$ as $|s| \to \infty$.

Denote the factors of the product (14) by $u(s, s_n)$. Then

$$\ln|P(s)| = \Sigma_1 + \Sigma_2,$$

where

$$\Sigma_1 = \sum_{|s/s_n| < 1/2} \ln|u(s, s_n)|, \quad \Sigma_2 = \sum_{|s/s_n| > 1/2} \ln|u(s, s_n)|.$$

Further, for the summands in Σ_1 we have

$$\ln|u(s, s_n)| \leq \frac{1}{p+1}\left|\frac{s}{s_n}\right|^{p+1} + \frac{1}{p+2}\left|\frac{s}{s_n}\right|^{p+2} + \ldots \leq 2\left|\frac{s}{s_n}\right|^{p+1}$$

and for the summands in Σ_2 we have

$$\ln|u(s, s_n)| \leq \ln\left(1 - \left|\frac{s}{s_n}\right|\right) + \left|\frac{s}{s_n}\right| + \ldots + \frac{1}{p}\left|\frac{s}{s_n}\right|^p$$

$$= r_p\left(\left|\frac{s}{s_n}\right|\right) \leq \begin{cases} c(p)\left|\dfrac{s}{s_n}\right|^p, & \text{if } p \geq 1; \\[3mm] c(\varepsilon)\left|\dfrac{s}{s_n}\right|^\varepsilon, & \text{if } p = 0. \end{cases}$$

Consequently,

$$\ln|P(s)| \ll \sum_{|s/s_n| \leq 1/2} \left|\frac{s}{s_n}\right|^{p+1} + \sum_{|s/s_n| > 1/2} r_p\left(\left|\frac{s}{s_n}\right|\right).$$

If $\beta = P + 1$, then the first sum is $\ll |s|^\beta$.
If $\beta < p + 1$ and $\beta + \varepsilon < p + 1$, then the first sum is

$$\ll |s|^{\beta+\varepsilon} \sum_{|s/s_n| < 1/2} \frac{1}{|s_n|^{\beta+\varepsilon}}\left|\frac{s}{s_n}\right|^{p+1-(\beta+\varepsilon)} \ll |s|^{\beta+\varepsilon}.$$

Thus, for any $p \geq 0$, the first sum is $\ll |s|^{\beta+\varepsilon}$. If $p \geq 1$, then the second sum (since

$\beta \geq p)$

$$\ll \sum_{|s/s_n| > 1/2} \left|\frac{s}{s_n}\right|^p = |s|^{\beta+\varepsilon} \sum_{|s/s_n| > 1/2} \frac{1}{|s_n|^{\beta+\varepsilon}} \cdot \left|\frac{s}{s_n}\right|^{p-(\beta+\varepsilon)} \ll |s|^{\beta+\varepsilon}.$$

If $p = 0$, the second sum

$$\ll \sum_{|s/s_n| > 1/2} \left|\frac{s}{s_n}\right|^\varepsilon = |s|^{\beta+\varepsilon} \sum_{|s/s_n| > 1/2} \frac{1}{|s_n|^{\beta+\varepsilon}} \left|\frac{s}{s_n}\right|^{-\beta} \ll |s|^{\beta+\varepsilon}.$$

This proves the first part of the Lemma. To prove the second part, we note that $\beta > 0$, since $|s_n| \to \infty$ and the series $\sum |s_n|^{-\beta}$ converges. Then, replacing $\beta + \varepsilon$ by β in the previous argument (i.e. replacing everywhere $|s|^\beta$) and taking $0 < \varepsilon < \beta$, we obtain the second assertion. This proves the Lemma. □

Now let $P(s)$ be an entire function of finite order α with $P(0) \neq 0$. By Theorem 4,

$$P(s) = e^{g(s)} \prod_{n=1}^{\infty} \left(1 - \frac{s}{s_n}\right) e^{\frac{s}{s_n} + \frac{1}{2}\left(\frac{s}{s_n}\right)^2 + \ldots + \frac{1}{n-1}\left(\frac{s}{s_n}\right)^{n-1}}.$$

By Lemma 2, the exponent of convergence of the sequence s_n does not exceed α. Let $p \geq 0$ be the smallest integer for which

$$\sum_{n=1}^{\infty} \frac{1}{|s_n|^{p+1}} < +\infty.$$

Then by Theorem 4

$$P(s) = e^{g(s)} \prod_{n=1}^{\infty} \left(1 - \frac{s}{s_n}\right) e^{\frac{s}{s_n} + \frac{1}{2}\left(\frac{s}{s_n}\right)^2 + \ldots + \frac{1}{p}\left(\frac{s}{s_n}\right)^p}, \tag{15}$$

where $g(s)$ is an entire function. It will be proved in Lemma 5 that $g(s)$ is a polynomial. For this we need the following Lemma, which is of independent interest, and will also be used in Chapter VI, §1.

Lemma 4. *Let $R > 0$, and let the function $f(s)$ defined by*

$$f(s) = \sum_{n=0}^{\infty} a_n(s - s_0)^n,$$

be analytic in the circle $|s - s_0| \leq R$. Moreover, let $\operatorname{Re} f(s) \leq M$ on the circle $|s - s_0| = R$. Then

a) $\dfrac{1}{n!}|f^{(n)}(s_0)| = |a_n| \leq 2\{M - \operatorname{Re} f(s_0)\} R^{-n}, \quad n \geq 1;$

b) in the disk $|s - s_0| \leq r < R$

$$|f(s) - f(s_0)| \leq 2\{M - \operatorname{Re} f(s_0)\}\frac{r}{R - r},$$

$$|f^{(n)}(s)| \leq 2n!\,\{M - \operatorname{Re} f(s_0)\}\frac{R}{(R - r)^{n+1}}, \quad n \geq 1.$$

Proof. We shall prove a) first for $s_0 = 0$ and $a_0 = f(0) = 0$. Since $\operatorname{Re} f(s)$ attains its maximum on the boundary, and $f(s) = 0$ for $s = 0$, then $M \geq 0$. Setting $a_n = |a_n| e^{i\varphi_n}$, $s = \operatorname{Re}^{i\varphi}$, we have

$$\operatorname{Re} f(\operatorname{Re}^{i\varphi}) = \sum_{n=1}^{\infty} |a_n| \cos(n\varphi + \varphi_n) R^n. \tag{16}$$

Since the series $\sum_{n=1}^{\infty} |a_n| R^n$ converges, it follows that the series (16) converges uniformly for $0 \leq \phi \leq 2\pi$, and so can be integrated term by term. This yields

$$\int_0^{2\pi} \operatorname{Re} f(\operatorname{Re}^{i\varphi}) d\varphi = 0.$$

Moreover,

$$\int_0^{2\pi} \operatorname{Re} f(\operatorname{Re}^{i\varphi}) \cos(n\varphi + \varphi_n) d\varphi = \pi |a_n| R^n, \quad n \geq 1.$$

Consequently, since $M \geq 0$ and $1 + \cos(n\phi + \phi_n) \geq 0$, we have

$$\pi |a_n| R^n = \int_0^{2\pi} \operatorname{Re} f(\operatorname{Re}^{i\varphi}) \{1 + \cos(n\varphi + \varphi_n)\} d\varphi \leq 2\pi M;$$

$$|a_n| \leq 2M R^{-n}.$$

If $s_0 \neq 0$, then we consider $F(s')$, where

$$F(s') = f(s' + s_0) - a_0 = a_1 s' + a_2 s'^2 + \dots$$

Then $F(0) = 0$ and $\operatorname{Re} F(s') \leq M - \operatorname{Re} f(s_0)$ for $|s'| = R$. This proves part a) of the Lemma.

b) Further,

$$|f(s) - f(s_0)| \leq 2\{M - \operatorname{Re} f(s_0)\} \sum_{n=1}^{\infty} \left(\frac{r}{R}\right)^n < 2\{M - \operatorname{Re} f(s_0)\} \frac{r}{R - r}$$

Differentiating term by term the series for $f(s)$, we find

$$|f^{(n)}(s)| \leq \sum_{m=n}^{\infty} |a_m| m(m-1) \dots (m-n+1)|s - s_0|^{m-n}$$

$$\leq 2\{M - \operatorname{Re} f(s_0)\} \sum_{m=n}^{\infty} m(m-1) \dots (m-n+1) R^{-m} r^{m-n} =$$

$$= 2\{M - \operatorname{Re} f(s_0)\} \frac{d^n}{dr^n} \sum_{m=0}^{\infty} \left(\frac{r}{R}\right)^m = 2n! \{M - \operatorname{Re} f(s_0)\} \frac{R}{(R-r)^{n+1}}.$$

This completes the proof of the Lemma. □

Lemma 5. *The entire function $g(s)$ in (15) is a polynomial of degree $g \leq \alpha$.*

Proof. Let $k = [\alpha]$. Then the number p in (15) does not exceed k. We shall prove that $g^{(k+1)}(s) \equiv 0$. Differentiating the logarithm of (15) $k + 1$ times gives

$$g^{(k+1)}(s) = \frac{d^k}{ds^k} \frac{P'(s)}{P(s)} + k! \sum_{n=1}^{\infty} \frac{1}{(s_n - s)^{k+1}}. \tag{17}$$

We consider the disk $|s| \le R/2$. For $|s_n| > R$,

$$|s_n - s| > \tfrac{1}{2}|s_n|,$$

Since the series $\Sigma |s_n|^{-(k+1)}$ converges, it follows that

$$k! \sum_{|s_n| > R} \frac{1}{|s_n - s|^{k+1}} < k! 2^{k+1} \sum_{|s_n| > R} |s_n|^{-(k+1)} \to 0 \tag{18}$$

as $R \to \infty$. We consider next the function

$$\frac{d^k}{ds^k} \left(\frac{P'(s)}{P(s)} + \sum_{|s_n| \le R} \frac{1}{s_n - s} \right),$$

which is the $(k + 1)$st derivative of

$$g_R(s) = \ln \left\{ \frac{P(s)}{P(0)} \prod_{|s_n| \le R} \left(1 - \frac{s}{s_n} \right)^{-1} \right\}.$$

The following estimate holds on the circle $|s| = 2R$:

$$\left| \frac{P(s)}{P(0)} \sum_{|s_n| \le R} \left(1 - \frac{s}{s_n} \right)^{-1} \right| < c(\varepsilon) e^{(2R)^{\alpha+\varepsilon}}.$$

By the maximum principle, this inequality also holds inside the disk $|s| \le 2R$, and so

$$\operatorname{Re} g_R(s) = \ln \left| \frac{P(s)}{P(0)} \prod_{|s_n| \le R} \left(1 - \frac{s}{s_n} \right)^{-1} \right| < c_1(\varepsilon) R^{\alpha+\varepsilon}$$

in the disk $|s| \le 2R$. Moreover, $\operatorname{Re} g_R(0) = 0$. Applying Lemma 4b) with $r = R/2$, we find

$$|g_R^{(k+1)}(s)| < 2(k + 1)! c_1(\varepsilon) R^{\alpha+\varepsilon} \frac{R}{(R/2)^{k+2}} < c_2(\varepsilon) R^{\alpha-(k+1)+\varepsilon}.$$

Since $\varepsilon > 0$ is arbitrarily small, then the right side of the last inequality tends to 0 as $R \to \infty$. From (17) and (18) we obtain that $g^{(k+1)}(s) \to 0$ as $|s| = R \to \infty$. i.e. $g^{(k+1)}(s) \equiv 0$. This proves the Lemma. $\qquad \square$

Proof of Theorem 5. Formula (15) follows from Theorem 4. Lemma 5 implies that $g(s)$ is a polynomial of degree $g \le \alpha$. Since the order of $e^{g(s)}$ equals g, and the order of the product over the zeros in (15) equals β, then the order α satisfies $\alpha \le \max(g, \beta)$. Further, if $\beta < \alpha$, then $\alpha = g$, and so

$$|G(s)| \le c(\varepsilon) e^{c_1 r^g} e^{c_2 r^{\beta+\varepsilon}},$$

where $c_1 > 0$ and $c_2 > 0$ are absolute constants, and $\varepsilon > 0$ is arbitrarily small. Thus, ε can be chosen so that $\beta + \varepsilon < \alpha$, which contradicts the hypothesis of the Theorem. Therefore, $\alpha = \beta$. The convergence of the series $\sum |s_n|^{-\beta}$ follows from Lemma 3. This completes the proof of Theorem 5. ☐

Corollary. *If $G_1(s)$ is an entire function of finite order α, then $G_1(s) = s^m G(s)$, where $G(s)$ is the entire function of Theorem 5 and $m \geq 0$.*

Exercises

1. Let $f(x)$ be a polynomial with real coefficients,

$$f(x) = \alpha_0 + \alpha_1 x + \ldots + \alpha_{n-1} x^{n-1},$$

let $A > 0$, and let $\mu = \mu(A, f)$ be the measure of the points x in the interval $[0, 1]$ for which $|f(x)| \leq A$. Prove that

$$\mu \leq \min\left(1, 4e(A\alpha^{-1})^{\frac{1}{n-1}}\right),$$

where

$$\alpha = \max |\alpha_j|, \quad j = 0, 1, \ldots, n - 1.$$

2. Let $f(x)$ be a polynomial with real coefficients

$$f(x) = \alpha_1 x + \ldots + \alpha_n x^n, \quad \alpha = \max |\alpha_j|, \quad j = 1, 2, \ldots, n.$$

Prove that

$$\left| \int_0^1 e^{2\pi i f(x)} \, dx \right| \leq \min(1, 32\alpha^{-1/n}).$$

3. Let $f(x_1, \ldots, x_r)$ be a polynomial with real coefficients,

$$f(x_1, \ldots, x_r) = \sum_{t_1=0}^{n} \ldots \sum_{t_r=0}^{n} \alpha(t_1, \ldots, t_r) x_1^{t_1} \ldots x_r^{t_r},$$

$$\alpha(0, \ldots, 0) = 0, \quad \alpha = \max_{0 \leq t_1, \ldots, t_r \leq n} |\alpha(t_1, \ldots, t_r)|.$$

Consider the multiple trigonometric integral I defined by

$$I = \int_0^1 \ldots \int_0^1 e^{2\pi i f(x_1, \ldots, x_r)} \, dx_1 \ldots dx_r.$$

Prove that

$$I \leq \min(1, 32^r \alpha^{-1/n} (\ln(\alpha + 1) + 1)^{r-1}).$$

4. Let $\alpha \geq 1, r \geq 1, n \geq 1$,

$$J = \int_0^1 \ldots \int_0^1 e^{2\pi i \alpha x_1^n \cdots x_r^n} dx_1 \ldots dx_r.$$

Then

$$|J| \geq \frac{1}{2\pi n^r (r-1)!} \alpha^{-1/n} (\ln \alpha)^{r-1}.$$

5. For $0 < x < 1$, let $f(x)$ be a real-valued function with n derivatives, $n > 1$, such that, for some $A > 0, f(x)$ satisfies

$$A \leq |f^{(n)}(x)|, \quad 0 < x < 1.$$

Then the measure U of the points x for which $|f'(x)| \leq B$ does not exceed

$$(2n - 2)(BA^{-1})^{\frac{1}{n-1}}.$$

6. Under the conditions of problem 5 for the integral J,

$$J = \int_0^1 e^{2\pi i f(x)} dx,$$

show that

$$|J| \leq \min(1, 6n A^{-1/n}).$$

7. Let $f(x)$ be a polynomial with real coefficients

$$f(x) = \alpha_1 x + \ldots + \alpha_n x^n,$$

$$\beta_r(x) = \frac{1}{r!} f^{(r)}(x), \quad r = 1, \ldots, n,$$

$$H = H(\alpha_1, \ldots, \alpha_n) = \min_{a \leq x \leq b} \sum_{r=1}^n |\beta_r(x)|^{1/r}.$$

Prove that the interval $a < x < b$ can be covered by m

$$m \leq \frac{n^2 + n}{2} - 1,$$

disjoint intervals such that on each of these intervals the inequality

$$|\beta_r(x)| \geq (n^{-1} H)^r.$$

is satisfied for some r, $1 \leq r \leq n$.

8. Under the conditions of problem 7 for the integral J,

$$J = \int_a^b e^{2\pi i f(x)} dx,$$

prove that

$$|J| \leq \min(b - a, 6en^3 H^{-1}).$$

9. Let $\theta = \theta(k)$ be a singular integral ("singular integral of Terry's problem")
of the form

$$\theta = \int\limits_{-\infty}^{+\infty} \ldots \int\limits_{-\infty}^{+\infty} \left| \int\limits_0^1 e^{2\pi i(\alpha_1 x + \ldots + \alpha_n x^n)} dx \right|^{2k} d\alpha_1 \ldots d\alpha_n.$$

Prove that θ converges for

$$2k > \frac{n^2 + n}{2} + 1.$$

Chapter III. The Euler Gamma Function

§1. Definition and Simplest Properties

The Euler gamma function $\Gamma(s)$ is defined by the equation

$$\frac{1}{\Gamma(s)} = se^{\gamma s} \prod_{n=1}^{\infty} \left(1 + \frac{s}{n}\right) e^{-s/n},$$

where γ is Euler's constant.

It follows from the definition and the theorems in Chapter 2 that $\Gamma^{-1}(s)$ is an entire function of order at most one. Moreover, $\Gamma(s)$ is an analytic function in the entire s-plane except for the points $s = 0, -1, -2, \ldots$, where it has simple poles.

Theorem 1 (Euler's formula).

$$\Gamma(s) = \frac{1}{s} \prod_{n=1}^{\infty} \left(1 + \frac{1}{n}\right)^s \left(1 + \frac{s}{n}\right)^{-1}. \tag{1}$$

Proof. From the definition of an infinite product (Ch. II, §1) and from the definition of the function $\Gamma(s)$, we obtain

$$\frac{1}{\Gamma(s)} = s \lim_{m \to \infty} e^{s(1 + \frac{1}{2} + \ldots + \frac{1}{m} - \log m)} \cdot \lim_{m \to \infty} \prod_{n=1}^{m} \left(1 + \frac{s}{n}\right) e^{-\frac{s}{n}}$$

$$= s \lim_{m \to \infty} m^{-s} \prod_{n=1}^{m} \left(1 + \frac{s}{n}\right) = s \lim_{m \to \infty} \prod_{n=1}^{m-1} \left(1 + \frac{1}{n}\right)^{-s} \prod_{n=1}^{m} \left(1 + \frac{s}{n}\right)$$

$$= s \lim_{m \to \infty} \prod_{n=1}^{m} \left(1 + \frac{1}{n}\right)^{-s} \left(1 + \frac{s}{n}\right) \left(1 + \frac{1}{m}\right)^s$$

$$= s \prod_{n=1}^{\infty} \left(1 + \frac{1}{n}\right)^{-s} \left(1 + \frac{s}{n}\right),$$

which is what we had to prove. □

Corollary 1.

$$\Gamma(s) = \lim_{n \to \infty} \frac{1 \cdot 2 \ldots (n-1)n^s}{s(s+1) \ldots (s+n-1)}.$$

Corollary 2. $\Gamma(1) = \Gamma(2) = 1$.

Theorem 2 (functional equation). $\Gamma(s+1) = s\Gamma(s)$.

Proof. We obtain from (1) that

$$\frac{\Gamma(s+1)}{\Gamma(s)} = \frac{s}{s+1} \lim_{m\to\infty} \prod_{n=1}^{m} \frac{\left(1+\dfrac{1}{n}\right)^{s+1}\left(1+\dfrac{s+1}{n}\right)^{-1}}{\left(1+\dfrac{1}{n}\right)^{s}\left(1+\dfrac{s}{n}\right)^{-1}}$$

$$= \frac{s}{s+1} \lim_{m\to\infty} \prod_{n=1}^{m} \frac{n+1}{n} \cdot \frac{n+s}{n+s+1} = \frac{s}{s+1} \lim_{m\to\infty} \frac{(m+1)(s+1)}{m+1+s} = s\,.$$

This proves the Theorem.   □

Corollary 1. $\Gamma(n+1) = n!$ *for every natural number n.*

Corollary 2 (duplication formula). *For every natural number n,*

$$\Gamma(2n)\Gamma\left(\frac{1}{2}\right) = 2^{2n-1}\Gamma(n)\Gamma\left(n+\frac{1}{2}\right).$$

Theorem 3 (addition formula). *For any number s not an integer,*

$$\Gamma(s)\Gamma(1-s) = \frac{\pi}{\sin \pi s}\,.$$

Proof. First we represent $\sin \pi s$ as an infinite product. The function $\sin \pi s$ is entire and of order one, and its zeros are at $s = 0, \pm 1, \pm 2, \ldots$, and so, by Theorem 5 of Chapter II

$$\sin \pi s = s e^{H(s)} \prod_{n=1}^{\infty}\left(1-\frac{s^2}{n^2}\right),$$

where $H(s) = as + b$.

Taking the logarithmic derivative of this equation, we find that

$$\pi \frac{\cos \pi s}{\sin \pi s} = \frac{1}{s} + H'(s) - \sum_{n=1}^{\infty} \frac{2s}{n^2 - s^2}\,.$$

Passage to the limit as $s \to 0$ gives $a = 0$, and so $H(s) = b$.

Thus,

$$\frac{\sin \pi s}{s} = c \prod_{n=1}^{\infty}\left(1-\frac{s^2}{n^2}\right).$$

Passing again to the limit as $s \to 0$ gives $c = \pi$, i.e.

$$\sin \pi s = \pi s \prod_{n=1}^{\infty}\left(1-\frac{s^2}{n^2}\right).$$

From the definition of the function $\Gamma(s)$, we have

$$\Gamma(s)\Gamma(-s) = -\frac{1}{s^2}\prod_{n=1}^{\infty}\left(1 - \frac{s^2}{n^2}\right)^{-1} = -\frac{\pi}{s\sin\pi s}.$$

Also, $\Gamma(1 - s) = -s\Gamma(-s)$ by Theorem 2. This proves the Theorem.   □

Corollary. $\Gamma(1/2) = \sqrt{\pi}$.

Theorem 4 (integral formula). *For* Re $s > 0$

$$\Gamma(s) = \int_0^{\infty} e^{-t}t^{s-1}\,dt.$$

Proof. Note that the integral on the right converges uniformly for Re $s \geq \sigma_0 > 0$, and, consequently, represents a function analytic in the half plane Re $s > 0$. From Corollary 1 to Theorem 1, we have

$$\Gamma(s) = \lim_{n\to\infty}\frac{1\cdot 2\ldots(n-1)}{s(s+1)\ldots(s+n-1)}n^s.$$

We consider the function

$$\Pi(s; n) = \int_0^n\left(1 - \frac{t}{n}\right)^n\cdot t^{s-1}\,dt = n^s\int_0^1(1-t)^n t^{s-1}\,dt$$

$$= \frac{n^s}{s}\int_0^1(1-t)^n\,dt^s = n^s\frac{n}{s}\int_0^1(1-t)^{n-1}t^s\,dt = \ldots$$

$$= n^s\frac{n(n-1)\ldots 1}{s(s+1)\ldots(s+n-1)}\int_0^1 t^{s+n-1}\,dt =$$

$$= n^s\frac{n(n-1)\ldots 1}{s(s+1)\ldots(s+n)}.$$

Consequently,

$$\Gamma(s) = \lim_{n\to\infty}\Pi(s; n) = \lim_{n\to\infty}\int_0^n\left(1 - \frac{t}{n}\right)^n t^{s-1}\,dt.$$

Let $\Gamma_1(s) = \int_0^{\infty} e^{-t}t^{s-1}\,dt$. Then

$$\Gamma_1(s) - \Gamma(s) = \lim_{n\to\infty}\int_0^n t^{s-1}\left(e^{-t} - \left(1 - \frac{t}{n}\right)^n\right)dt.$$

Note that $\left(1 + \frac{t}{n}\right)^n \leq e^t$ for $|t| < n$. Moreover, for $0 < y < 1$

$$1 - ny \leq (1 - y)^n.$$

Therefore,

$$0 \le e^{-t} - \left(1 - \frac{t}{n}\right)^n = e^{-t}\left(1 - e^t\left(1 - \frac{t}{n}\right)^n\right)$$

$$\le e^{-t}\left(1 - \left(1 - \frac{t^2}{n^2}\right)^n\right) \le e^{-t}\frac{t^2}{n};$$

$$\left|\int_0^n t^{s-1}\left(e^{-t} - \left(1 - \frac{t}{n}\right)^n\right)dt\right| \le \frac{1}{n}\int_0^\infty t^{\sigma+1}e^{-t}\,dt$$

and so $\Gamma_1(s) = \Gamma(s)$, which is what had to be proved.   □

Corollary.

$$\int_{-\infty}^{+\infty} e^{-u^2}\,du = \sqrt{\pi}.$$

§2. Stirling's Formula

It is important in applications to know the behavior of $\Gamma(s)$ as $|s| \to \infty$.

Theorem 5 (Stirling's formula). *For $\delta > 0$ and $-\pi + \delta \le \arg s \le \pi - \delta$ we have*

$$\log \Gamma(s) = \left(s - \frac{1}{2}\right)\log s - s + \log\sqrt{2\pi} + O\left(\frac{1}{|s|}\right),$$

where the constant in the symbol O depends only on δ.

Proof. From the definition of $\Gamma(s)$ we find that

$$\log \Gamma(s) = -\gamma s - \log s + \sum_{n=1}^{\infty}\left(\frac{s}{n} - \log\left(1 + \frac{s}{n}\right)\right). \tag{2}$$

For the natural number N we consider the two sums

$$\Sigma_1 = \sum_{1/2 < n \le N+1/2} 1/n, \quad \Sigma_2 = \sum_{1/2 < n \le N+1/2} \log(n+s).$$

Applying Theorem 1 of Chapter I to each sum, we find that

$$\Sigma_1 = \log N + \gamma + O\left(\frac{1}{N}\right);$$

$$\Sigma_2 = (N + \tfrac{1}{2} + s)\log(N + \tfrac{1}{2} + s) - (N + \tfrac{1}{2} + s)$$
$$- (s + \tfrac{1}{2})\log(s + \tfrac{1}{2}) + (s + \tfrac{1}{2})$$
$$+ \frac{\sigma(\tfrac{1}{2})}{s + \tfrac{1}{2}} - \frac{\sigma(N + \tfrac{1}{2})}{N + \tfrac{1}{2} + s} - \int_{1/2}^{N+1/2} \frac{\sigma(x)\,dx}{(x+s)^2}.$$

It follows from these formulas for Σ_1 and Σ_2 and from (2) that

$$\log \Gamma(s) = (s - \tfrac{1}{2}) \log s - s + c + J, \qquad (3)$$

where c is an absolute constant, and

$$J = \int_{1/2}^{+\infty} \frac{\sigma(x)\,dx}{(x+s)^2}.$$

Further,

$$J = O\left(\int_{1/2}^{+\infty} \frac{dx}{x^2 + |s|^2 - 2x|s|\cos\delta} \right) = O\left(\frac{1}{|s|} \right).$$

Thus, we have obtained

$$\log \Gamma(s) = (s - \tfrac{1}{2}) \log s - s + c + O\left(\frac{1}{|s|} \right).$$

We evaluate this formula at $s = n$, $n + 1/2$, and $2n$, and then apply the duplication formula and the fact that $\Gamma(1/2) = \sqrt{\pi}$. As $n \to \infty$ we obtain $c = \log \sqrt{2\pi}$. This completes the proof. $\qquad\square$

Corollary 1. *The function $\Gamma^{-1}(s)$ is an entire function of order one.*

Corollary 2. *For $\alpha \leq \sigma \leq \beta$ and $t \to \infty$,*

$$\Gamma(\sigma + it) = t^{\sigma + it - \frac{1}{2}} e^{-\frac{\pi t}{2} - it + i\frac{\pi}{2}(\sigma - \frac{1}{2})} \sqrt{2\pi}\{1 + O(1/t)\},$$

where the constant in the symbol O depends only on α and β.

Corollary 3. *Differentiating (3), we find for $|\arg s| < \pi$*

$$\frac{\Gamma'(s)}{\Gamma(s)} = \log s + O\left(\frac{1}{|s|} \right).$$

§3. The Euler Beta Function and Dirichlet's Integral

Closely related to the gamma function are the Euler beta function and Dirichlet's integral. We shall derive some of their basic properties.

Definition. For Re $u > 0$ and Re $v > 0$, the *Euler beta function* is defined by the equation

$$B(u, v) = \int_0^1 x^{u-1}(1 - x)^{v-1}\,dx.$$

Lemma. *The Euler beta function satisfies the identity*

$$B(u, v) = \frac{\Gamma(u)\Gamma(v)}{\Gamma(u + v)}.$$

Proof. Without loss of generality, we can assume that $\sigma_1 = \mathrm{Re}\ u > 1$ and $\sigma_2 = \mathrm{Re}\ v > 1$. Moreover, for $Y \geq 0$

$$\left| \int_0^Y t^{u-1} e^{-t}\, dt \right| \leq \int_0^Y t^{\sigma_1 - 1} e^{-t}\, dt = O(\min(1, Y)),$$

and for $X \geq 1$

$$\left| \int_X^\infty t^{u-1} e^{-t}\, dt \right| \leq \int_X^\infty t^{\sigma_1 - 1} e^{-t}\, dt = O(X^{\sigma_1 - 1} e^{-X}),$$

where the constant in the symbol O depends only on σ_1. Let $X > 1$ and $Y = X^{-0.5}$. Then

$$\Gamma(u) = \int_Y^X t^{u-1} e^{-t}\, dt + O(Y); \quad \Gamma(v) = \int_0^{X^2} \tau^{v-1} e^{-\tau}\, d\tau + O(Y).$$

Multiplying these relations, we obtain

$$\Gamma(u)\Gamma(v) = J + O(Y),$$

where

$$J = \int_Y^X t^{u-1} e^{-t} \left(\int_0^{X^2} \tau^{v-1} e^{-\tau}\, d\tau \right) dt.$$

In the inner integral we make a change of variables of the form $\tau = zt$, $0 \leq z \leq X^2 t^{-1}$. This gives

$$J = \int_Y^X t^{u+v-1} e^{-t} \left(\int_0^{X^2 t - 1} z^{v-1} e^{-tz}\, dz \right) dt$$

$$= \int_Y^X t^{u+v-1} e^{-t} \left(\int_0^X z^{v-1} e^{-tz}\, dz \right) dt + O(e^{-\sqrt{X}}) = J_1 + O(e^{-\sqrt{X}}),$$

where

$$J_1 = \int_0^X z^{v-1} \left(\int_Y^X t^{u+v-1} e^{-t(z+1)}\, dt \right) dz.$$

In the inner integral we again make a change of variables of the form $t(z + 1) = x$, $(z + 1)Y \leq x \leq (z + 1)X$. We obtain

$$J_1 = \int_0^X \frac{z^{v-1}}{(z+1)^{u+v}} \left(\int_{(z+1)Y}^{(z+1)X} x^{u+v-1} e^{-x}\, dx \right) dz =$$

$$= \int_0^{\sqrt[4]{X}} \frac{z^{v-1}}{(z+1)^{u+v}} \left(\int_{(z+1)Y}^{(z+1)X} x^{u+v-1} e^{-x}\, dx \right) dz + R,$$

where

$$|R| \le \int_{\sqrt[4]{X}}^{X} \frac{z^{\sigma_2-1}\Gamma(\sigma_1 + \sigma_2)}{(z+1)^{\sigma_1+\sigma_2}} \, dz = O(Y^{0,5\sigma_1}).$$

In addition, for $0 \le z \le \sqrt[4]{X} = Y^{-0,5}$, we have

$$\int_{(z+1)Y}^{(z+1)X} x^{u+v-1} e^{-x} \, dx = \Gamma(u+v) + O(\sqrt{Y}).$$

Consequently,

$$J_1 = \Gamma(u+v) \int_0^{\sqrt[4]{X}} \frac{z^{v-1}}{(z+1)^{u+v}} \, dz + O(\sqrt{Y}) + O(Y^{0,5\sigma_1}).$$

Finally,

$$\int_0^{\sqrt[4]{X}} \frac{z^{v-1}}{(z+1)^{u+v}} \, dz = \int_0^{+\infty} \frac{z^{v-1}}{(z+1)^{u+v}} \, dz + O(Y^{0,5\sigma_1})$$

$$= 2 \int_0^{\pi/2} (\cos\theta)^{2u-1} (\sin\theta)^{2v-1} \, d\theta + O(Y^{0,5\sigma_1})$$

$$= \int_0^1 x^{u-1}(1-x)^{v-1} \, dx + O(Y^{0,5\sigma_1})$$

$$= B(u,v) + O(Y^{0,5\sigma_1}).$$

Thus, we have obtained the relation

$$\Gamma(u)\Gamma(v) = \Gamma(u+v)B(u,v) + O(\sqrt{Y}) + O(Y^{0,5\sigma_1}).$$

The Lemma now follows by passing to the limit in this equation as $X \to +\infty$.
□

Theorem 6. *Let $f(u)$ be a continuous function, let $\alpha_1 > 0, \alpha_2 > 0, \ldots, \alpha_n > 0$, and let*

$$I = \int\cdots\int_{\substack{t_1+t_2+\ldots+t_n \le 1 \\ 0 \le t_1,t_2,\ldots,t_n \le 1}} f(t_1 + t_2 + \ldots + t_n) t_1^{\alpha_1-1} t_2^{\alpha_2-1} \cdots$$

$$t_n^{\alpha_n-1} \, dt_1 \, dt_2 \ldots dt_n.$$

Then

$$I = \frac{\Gamma(\alpha_1)\Gamma(\alpha_2)\ldots\Gamma(\alpha_n)}{\Gamma(\alpha_1+\alpha_2+\ldots+\alpha_n)} \int_0^1 f(u)u^{\alpha_1+\alpha_2+\ldots+\alpha_n-1} \, du.$$

Proof. Let $\lambda = t_3 + \ldots + t_n$ and consider the integral

$$I_1 = \int_0^{1-\lambda} \left(\int_0^{1-\lambda-t_2} f(t_1 + t_2 + \lambda) t_1^{\alpha_1-1} \, dt_1 \right) t_2^{\alpha_2-1} \, dt_2.$$

In I_1 we make the change of variables $t_1 = t_2(1 - v)/v$, and then change the order of integration:

$$I_1 = \int_0^{1-\lambda} \left(\int_{t_2/1-\lambda}^1 f\left(\lambda + \frac{t_2}{v} \right)(1 - v)^{\alpha_1 - 1} v^{-\alpha_1 - 1} \, dv \right) t_2^{\alpha_1 + \alpha_2 - 1} \, dt_2$$

$$= \int_0^1 \left(\int_0^{(1-\lambda)v} f\left(\lambda + \frac{t_2}{v} \right) t_2^{\alpha_1 + \alpha_2 - 1} \, dt_2 \right)(1 - v)^{\alpha_1 - 1} v^{-\alpha_1 - 1} \, dv.$$

In the inner integral we make the change of variable $t_2 = \tau v$:

$$I_1 = \int_0^1 \left(\int_0^{1-\lambda} f(\lambda + \tau)\tau^{\alpha_1 + \alpha_2 - 1} \, d\tau \right)(1 - v)^{\alpha_1 - 1} v^{\alpha_2 - 1} \, dv$$

$$= \frac{\Gamma(\alpha_1)\Gamma(\alpha_2)}{\Gamma(\alpha_1 + \alpha_2)} \int_0^{1-\lambda} f(\lambda + \tau)\tau^{\alpha_1 + \alpha_2 - 1} \, d\tau.$$

For I we have the formula

$$I = \frac{\Gamma(\alpha_1)\Gamma(\alpha_2)}{\Gamma(\alpha_1 + \alpha_2)} \int_{\substack{\tau + t_3 + \ldots + t_n \leqslant 1 \\ 0 \leqslant \tau, t_3, \ldots, t_n \leqslant 1}} \cdots \int f(\tau + t_3 + \ldots + t_n)$$

$$\times \tau^{\alpha_1 + \alpha_2 - 1} t_3^{\alpha_3 - 1} \ldots t_n^{\alpha_n - 1} \, d\tau \, dt_3 \ldots dt_n.$$

This implies the Theorem.   □

The integral I in Theorem 6 is called *Dirichlet's integral*.

Exercises

1. Let M and N be natural numbers, $M \geq 2$,

$$I(n) = \int_0^1 \frac{\sin \pi(2M + 1)u}{\sin \pi u} e^{2\pi i \frac{(n+u)^2}{N}} \, du.$$

Prove that

$$I(n) = e^{2\pi i \frac{n^2}{N}} + O\left(\frac{\ln M}{M} \right).$$

2. Under the conditions of exercise 1, show that

$$S = \sum_{n=0}^{N-1} e^{2\pi i \frac{n^2}{N}} = \sum_{h=-M}^M J(k) + O\left(\frac{N \ln M}{M} \right),$$

where

$$J(k) = Ne^{-2\pi i \frac{Nk^2}{4}} \int_{0,5k}^{0,5k+1} e^{2\pi iNu^2}\, du.$$

3. With the notation of exercise 2, prove that

$$\sum_{k=-2M}^{2M} J(k) = N \int_{-M}^{M+1} e^{2\pi iNu^2}\, du + Ne^{-2\pi i \frac{N}{4}} \int_{-M+0,5}^{M+0,5} e^{2\pi iNu^2}\, du.$$

4. Prove the identity ("Gauss's sum and its argument"):

$$S = \sum_{n=0}^{N-1} e^{2\pi i \frac{n^2}{N}} = \frac{1 + i^{-N}}{1 + i^{-1}} \sqrt{N}.$$

5. Let p be a prime number that divides the product of the two natural numbers n and m. Using the method of mathematical induction (by induction over p), prove that p divides either n or m.

6. Prove that if

$$n = p_1^{\alpha_1} \ldots p_r^{\alpha_r} = q_1^{\beta_1} \ldots q_s^{\beta_s},$$

where $p_1, \ldots, p_r, q_1, \ldots, q_r$ are prime numbers such that $p_1 < \ldots < p_r$ and $q_1 < \ldots < q_s$, and $\alpha_1, \ldots, \alpha_r, \beta_1, \ldots, \beta_s$ are natural numbers, then

$$r = s, \quad p_1 = q_1, \quad \alpha_1 = \beta_1, \ldots, p_r = q_r, \quad \alpha_r = \beta_r$$

This result on the unique decomposition of a natural number into prime factors is usually called the Fundamental Theorem of Arithmetic.

7. a) Prove that if $a_0, a_1, \ldots, a_k$ are integers, and if m is a natural number such that $m > k!$, and if

$$a_0 \ln m + a_1 \ln(m + 1) + \ldots + a_k \ln(m + k) = 0,$$

then

$$a_0 = a_1 = \ldots = a_k = 0.$$

b) Prove that for any natural number m and $k \geq 8\sqrt[3]{m}$ there are integers $a_0, a_1, \ldots, a_k$ not all equal to zero such that

$$a_0 \ln m + a_1 \ln(m + 1) + \ldots + a_k \ln(m + k) = 0.$$

8. Prove that the following inequalities hold for any natural number k:

a) $\displaystyle\sum_{n \leq X} \tau_k(n) \leq \frac{1}{(k-1)!} X(\ln X + k - 1)^{k-1}$;

b) $\displaystyle\sum_{n \leq X} \tau_k^2(n) \leq k^2(k!)^{-(k+1)} X(\ln X + k^2 - 1)^{k^2-1}.$

9. Let $1 \leq u < N$. Prove that the following identity holds for any complex-valued function $f(x)$:

$$\sum_{u < n \leq N} \Lambda(n) f(n) = \sum_{d \leq u} \mu(d) \sum_{l \leq Nd^{-1}} (\log l) f(ld)$$

$$- \sum_{d \leq u} \mu(d) \sum_{n \leq u} \Lambda(n) \sum_{r \leq N(dn)^{-1}} f(ndr)$$

$$- \sum_{u < m \leq Nu^{-1}} \left(\sum_{\substack{d \backslash m \\ d \leq u}} \mu(d) \right) \sum_{u < n \leq Nm^{-1}} \Lambda(n) f(nm).$$

Chapter IV. The Riemann Zeta Function

§1. Definition and Simplest Properties

Definition. For Re $s = \sigma > 1$, the *Riemann zeta function* $\zeta(s)$ is defined by

$$\zeta(s) = \sum_{n=1}^{\infty} \frac{1}{n^s}.$$

It follows from the definition that $\zeta(s)$ is an analytic function in the half-plane Re $s > 1$.

Lemma 1 (the Euler product). *For Re $s > 1$,*

$$\zeta(s) = \prod_p \left(1 - \frac{1}{p^s}\right)^{-1}.$$

Proof. Let Re $s > 1$ and let $X \geq 2$ be an integer. It follows from the absolute convergence of the series $1 + \dfrac{1}{p^s} + \dfrac{1}{p^{2s}} + \ldots$ and from the unique decomposition of a natural number into prime factors that

$$\prod_{p \leq X} \left(1 - \frac{1}{p^s}\right)^{-1} = \prod_{p \leq X} \left(1 + \frac{1}{p^s} + \frac{1}{p^{2s}} + \ldots\right) = \sum_{n \leq X} \frac{1}{n^s} + R(s; X),$$

where

$$|R(s; X)| \leq \sum_{n > X} \left|\frac{1}{n^s}\right| = \sum_{n > X} \frac{1}{n^\sigma} \leq \frac{1}{\sigma - 1} X^{1-\sigma}.$$

The Lemma follows by taking the limit as $X \to +\infty$. $\qquad \square$

Corollary. $\zeta(s) \neq 0$ *for Re $s > 1$.*
Indeed, for Re $s = \sigma > 1$,

$$\frac{1}{|\zeta(s)|} = \left|\prod_p \left(1 - \frac{1}{p^s}\right)\right| \leq \prod_p \left(1 + \frac{1}{p^\sigma}\right) < \sum_{n=1}^{\infty} \frac{1}{n^\sigma} \leq 1 + \int_1^\infty \frac{du}{u^\sigma} = \frac{\sigma}{\sigma - 1};$$

$$|\zeta(s)| > \frac{\sigma - 1}{\sigma} > 0. \qquad \square$$

We shall extend $\zeta(s)$ to the half-plane Re $s > 0$.

Lemma 2. *For Re $s > 0$ and $N \geq 1$, we have*

$$\zeta(s) = \sum_{n=1}^{N} \frac{1}{n^s} + \frac{N^{1-s}}{s-1} - \frac{1}{2}N^{-s} + s\int_{N}^{\infty} \frac{\rho(u)}{u^{s+1}}\, du,$$

where

$$\rho(u) = \frac{1}{2} - \{u\}.$$

Proof. Choose a natural number $M > N$. Applying formula (2) of Chapter I, we obtain

$$\sum_{N+\frac{1}{2} < n \leq M+\frac{1}{2}} \frac{1}{n^s} = \int_{N+\frac{1}{2}}^{M+\frac{1}{2}} \frac{du}{u^s} + s\int_{N+\frac{1}{2}}^{M+\frac{1}{2}} \frac{\rho(u)}{u^{s+1}}\, du$$

$$= \frac{1}{1-s}\left(M+\frac{1}{2}\right)^{1-s} + \frac{1}{s-1}N^{1-s} - \frac{1}{2}N^{-s} + s\int_{N}^{M+\frac{1}{2}} \frac{\rho(u)}{u^{s+1}}\, du.$$

Consequently, for Re $s > 1$

$$\zeta(s) = \sum_{n=1}^{N} \frac{1}{n^s} + \frac{N^{1-s}}{s-1} - \frac{1}{2}N^{-s} + s\int_{N}^{\infty} \frac{\rho(u)}{u^{s+1}}\, du.$$

The last integral defines an analytic function in the half-plane Re $s > 0$. The Lemma now follows by analytic continuation. □

Corollary. $\zeta(s)$ *is an analytic function in the half-plane Re $s > 0$ except at the point $s = 1$. At the point $s = 1$ the function $\zeta(s)$ has a simple pole with residue equal to 1.*

Before analytically continuing $\zeta(s)$ to the entire s-plane, we shall prove the following Lemma.

Lemma 3. *Let $x > 0$ and let α be a real number. Define $\theta(x, \alpha)$ by*

$$\theta(x, \alpha) = \sum_{n=-\infty}^{+\infty} e^{-\pi x(n+\alpha)^2}.$$

Then

$$\theta\left(\frac{1}{x}, \alpha\right) = \sqrt{x} \sum_{n=-\infty}^{+\infty} e^{-\pi n^2 x + 2\pi i n \alpha}.$$

Proof. Without loss of generality, we can assume that $0 \leq \alpha < 1$. Choose $N > 10$, $M = N^5$, and consider the integral

$$I(n) = \int_{-0.5}^{+0.5} \frac{\sin \pi(2M+1)u}{\sin \pi u}\, e^{-\pi x(n+\alpha+u)^2}\, du.$$

Since

$$\int_{-0.5}^{+0.5} \frac{\sin \pi (2M+1)u}{\sin \pi u}\, du = \sum_{k=-M}^{+M} \int_{-0.5}^{+0.5} e^{-2\pi i k u}\, du = 1,$$

then

$$I(n) = e^{-\pi x (n+\alpha)^2} + R(n), \tag{1}$$

where

$$R(n) = \int_{-0.5}^{+0.5} \frac{\sin \pi (2M+1)u}{\sin \pi u}(e^{-\pi x(n+\alpha+u)^2} - e^{-\pi x(n+\alpha)^2})\, du.$$

We shall estimate $|R(n)|$ under the condition that $-N \le n \le N$. We have

$$R(n) = I_1 + I_2 + I_3,$$

where

$$I_1 = \int_{-0.5}^{-N^{-3}} \Phi(u)\, du, \quad I_2 = \int_{-N^{-3}}^{+N^{-3}} \Phi(u)\, du, \quad I_3 = \int_{+N^{-3}}^{+0.5} \Phi(u)\, du,$$

$$\Phi(u) = \frac{\sin \pi (2M+1)u}{\sin \pi u}(e^{-\pi x(n+\alpha+u)^2} - e^{-\pi x(n+\alpha)^2}).$$

Let us estimate I_2. For this purpose, we estimate $|\Phi(u)|$ for $|u| \le N^{-3}$, using the formula for finite increments

$$|\Phi(u)| = O\left(\frac{|u|N}{|\sin \pi u|}\right) = O(N).$$

Consequently,

$$I_2 = O\left(\frac{1}{N^2}\right).$$

The integrals I_1 and I_3 are estimated in the same way. To estimate I_3, we integrate by parts and obtain

$$I_3 = \frac{-\cos \pi (2M+1)u}{\pi(2M+1)\sin \pi u}(e^{-\pi x(n+\alpha+u)^2} - e^{-\pi x(n+\alpha)^2})\Big|_{N^{-3}}^{0.5}$$

$$+ \int_{N^{-3}}^{0.5} \frac{\cos \pi (2M+1)u}{\pi(2M+1)} Y(u)\, du,$$

where

$$Y(u) = \frac{d}{du}\left(\frac{e^{-\pi x(n+\alpha+u)^2} - e^{-\pi x(n+\alpha)^2}}{\sin \pi u}\right).$$

A rough estimate for $|Y(u)|$ for $N^{-3} \le u \le 0.5$ gives

$$Y(u) = O\left(\frac{1}{u^2}\right) + O\left(\frac{N}{u}\right).$$

Consequently,

$$I_3 = O\left(\frac{N^3}{M}\right) + O\left(\frac{N \ln N}{M}\right) = O\left(\frac{1}{N^2}\right).$$

Thus,

$$R(n) = O\left(\frac{1}{N^2}\right).$$

Summing (1) over n for $-N \le n \le N$, we obtain

$$\sum_{n=-N}^{N} e^{-\pi x(n+\alpha)^2} + O\left(\frac{1}{N}\right) = \sum_{n=-N}^{+N} \int_{-0.5}^{+0.5} \sum_{k=-M}^{+M} e^{-2\pi i k u - \pi x(n+\alpha+u)^2} \, du$$

$$= \sum_{k=-M}^{+M} \sum_{n=-N}^{+N} \int_{-0.5}^{0.5} e^{-2\pi i k(n+u) - \pi x(n+u+\alpha)^2} \, du$$

$$= \sum_{k=-M}^{+M} \int_{-N-0.5}^{N+0.5} e^{-2\pi i k u - \pi x(u+\alpha)^2} \, du = \sum_{k=-M}^{+M} J(k) \, du.$$

Further,

$$J(k) = \int_{-N-0.5}^{N+0.5} e^{-2\pi i k u - \pi x(u+\alpha)^2} \, du = \int_{-\infty}^{+\infty} e^{-2\pi i k u - \pi x(u+\alpha)^2} \, du$$

$$+ O(e^{-\pi x N}) = e^{2\pi i k \alpha} J(k; x) + O(e^{-\pi x N}),$$

where

$$J(k; x) = \int_{-\infty}^{+\infty} e^{-2\pi i k u - \pi x u^2} \, du.$$

We next compute $J(k; 1/x)$. We have

$$J\left(k; \frac{1}{x}\right) = x \int_{-\infty}^{+\infty} e^{-2\pi i k u x - \pi x u^2} \, du = x e^{-\pi x k^2} \int_{-\infty}^{+\infty} e^{-\pi x(u+ki)^2} \, du.$$

Let $X > 1$ and let Γ be the contour of the rectangle with vertices at $-X$, $+X$, $-X + ik$, $X + ik$. Then

$$0 = \int_{\Gamma} e^{-\pi x u^2} \, du = \int_{-X}^{+X} e^{-\pi x u^2} \, du - \int_{-X}^{+X} e^{-\pi x(u+ki)^2} \, du + i \int_{0}^{k} e^{-\pi x/(X+iu)^2} \, du$$

$$- i \int_{0}^{k} e^{-\pi x(-X+iu)^2} \, du. \qquad (2)$$

The absolute values of the last two integrals do not exceed

$$\int_{0}^{k} e^{-\pi x X^2} e^{+\pi x u^2} \, du \le k e^{\pi x k^2} e^{-\pi x X^2}.$$

Taking the limit in (2) as $X \rightarrow +\infty$ gives

$$\int\limits_{-\infty}^{+\infty} e^{-\pi x(u+ki)^2}\, du = \int\limits_{-\infty}^{+\infty} e^{-\pi x u^2}\, du = \frac{1}{\sqrt{\pi x}} \int\limits_0^\infty e^{-v} v^{-\frac{1}{2}}\, dv = \frac{1}{\sqrt{\pi x}} \Gamma\left(\frac{1}{2}\right) = \frac{1}{\sqrt{x}}.$$

Thus,

$$J\left(k; \frac{1}{x}\right) = e^{-\pi x k^2} \sqrt{x};$$

$$\sum_{n=-N}^{N} e^{-\pi \frac{1}{x}(n+\alpha)^2} + O\left(\frac{1}{N}\right) = \sqrt{x} \sum_{k=-M}^{+M} e^{-\pi x k^2 + 2\pi i k\alpha} + O(e^{-\pi\frac{1}{x}N}).$$

Taking the limit in the last expression as $N \rightarrow +\infty$ gives the assertion of the Lemma. □

Corollary 1. *Let Re $s > 0$, let α be a real number, and define $\theta(s, \alpha)$ by*

$$\theta(s, \alpha) = \sum_{n=-\infty}^{+\infty} e^{-\pi s(n+\alpha)^2}.$$

Then

$$\theta\left(\frac{1}{s}, \alpha\right) = \sqrt{s} \sum_{n=-\infty}^{+\infty} e^{-\pi n^2 s + 2\pi i n\alpha}.$$

Corollary 2.

$$\theta\left(\frac{1}{x}, 0\right) = \sqrt{x}\,\theta(x, 0).$$

Theorem 1 (functional equation of the zeta function).

$$\pi^{-s/2} \Gamma\left(\frac{s}{2}\right) \zeta(s) = \pi^{-(1-s)/2} \Gamma\left(\frac{1-s}{2}\right) \zeta(1-s).$$

Proof. By Theorem 4 of Chapter III, we have for Re $s > 0$ and any natural number n

$$\Gamma\left(\frac{s}{2}\right) = \int\limits_0^\infty e^{-u} u^{\frac{s}{2}-1}\, du = n^s \int\limits_0^\infty e^{-\pi n^2 x} \pi^{\frac{s}{2}} x^{\frac{s}{2}-1}\, dx,$$

i.e.

$$\pi^{-\frac{s}{2}} \Gamma\left(\frac{s}{2}\right) n^{-s} = \int\limits_0^\infty e^{-\pi n^2 x} x^{\frac{s}{2}-1}\, dx.$$

Consequently, for Re $s > 1$

$$\pi^{-\frac{s}{2}} \Gamma\left(\frac{s}{2}\right) \zeta(s) = \int\limits_0^\infty x^{\frac{s}{2}-1} \left(\sum_{n=1}^\infty e^{-\pi n^2 x}\right) dx. \tag{3}$$

We can change the order of summation and integration since

$$\sum_{n>N} e^{-\pi n^2 x} = O(e^{-\pi N^2 x}),$$

for $x \geq 1$ and

$$\sum_{n=1}^{\infty} e^{-\pi n^2 x} = O\left(\frac{1}{\sqrt{x}}\right)$$

for $0 < x < 1$. Furthermore, if $\omega(x) = \sum_{n=1}^{\infty} e^{-\pi n^2 x}$, then Corollary 2 of Lemma 3 implies that

$$\omega\left(\frac{1}{x}\right) = -\frac{1}{2} + \frac{1}{2}x^{\frac{1}{2}} + x^{\frac{1}{2}}\omega(x).$$

Therefore,

$$\int_0^{\infty} x^{\frac{s}{2}-1}\omega(x)\,dx = \int_0^1 x^{\frac{s}{2}-1}\omega(x)\,dx + \int_1^{\infty} x^{\frac{s}{2}-1}\omega(x)\,dx$$

$$= \int_1^{\infty}\left(x^{-\frac{s}{2}-1}\omega\left(\frac{1}{x}\right) + x^{\frac{s}{2}-1}\omega(x)\right)dx$$

$$= \frac{1}{s(s-1)} + \int_1^{\infty}\left(x^{\frac{s}{2}-1} + x^{-\frac{s}{2}-\frac{1}{2}}\right)\omega(x)\,dx. \qquad (4)$$

Since $\omega(x) = O(e^{-\pi x})$ as $x \to +\infty$, it follows from (4) that the right side of (3) is an analytic function for any $s \neq 0, 1$ and is invariant under the interchange of s and $1 - s$, i.e.

$$\pi^{-\frac{s}{2}}\Gamma\left(\frac{s}{2}\right)\zeta(s) = \pi^{-\frac{1-s}{2}}\Gamma\left(\frac{1-s}{2}\right)\zeta(1-s). \qquad \square$$

Corollary. *The function*

$$\xi(s) = \frac{1}{2}s(s-1)\pi^{-\frac{s}{2}}\Gamma\left(\frac{s}{2}\right)\zeta(s)$$

is entire and

$$\xi(s) = \xi(1-s).$$

§2. Simplest Theorems on the Zeros

We see from Theorem 1 that the zeta function is equal to 0 at $s = -2$, $-4, \ldots, -2n, \ldots$, since $\Gamma^{-1}(s/2) = 0$ at these points. At $s = 0$ the zeta function is not equal to 0, since the zero of $\Gamma^{-1}(s/2)$ cancels the pole of $\zeta(1-s)$.

These zeros are called the *trivial zeros*. In addition to the trivial zeros, the zeta function has infinitely many *nontrivial zeros* lying in the strip $0 \leq \mathrm{Re}\ s \leq 1$ (*the critical strip*).

Theorem 2. *The function $\xi(s)$ is an entire function of order one that has infinitely many zeros ρ_n such that $0 \leq \mathrm{Re}\ \rho_n \leq 1$. The series $\sum |\rho_n|^{-1}$ diverges, but the series $\sum |\rho_n|^{-1-\varepsilon}$ converges for any $\varepsilon > 0$. The zeros of $\xi(s)$ are the nontrivial zeros of $\zeta(s)$.*

Proof. For $\mathrm{Re}\ s > 1$, the zeta function, and, consequently, $\xi(s)$, have no zeros. It follows from Theorem 1 that $\xi(s) \neq 0$ for $\mathrm{Re}\ s < 0$. Since $\xi(0) = \xi(1) \neq 0$, the zeros of $\xi(s)$ are precisely the nontrivial zeros of $\zeta(s)$.

We shall determine the order of $\xi(s)$. To do this we estimate $\xi(s)$ as $|s| \to \infty$. It suffices to do this only for $\mathrm{Re}\ s \geq 1/2$. Lemma 2 implies that $\zeta(s) = O(|s|)$ for $\mathrm{Re}\ s \geq 1/2$. Since $|\Gamma(s)| \leq e^{c|s|\ln|s|}$, the order of $\xi(s)$ is not greater than one. But $\ln \Gamma(s) \sim s \ln s$ as $s \to \infty$, and so the order of $\xi(s)$ equals 1. It follows from Theorem 5 of Chapter II that the series $\sum |\rho_n|^{-1}$, where the ρ_n are the zeros of $\xi(s)$, diverges, and, consequently, $\xi(s)$ has infinitely many zeros, and the series $\sum |\rho_n|^{-1-\varepsilon}$ converges for any $\varepsilon > 0$. This proves the Theorem. $\qquad \square$

Corollary 1. *We have the formula*

$$\xi(s) = e^{A+Bs} \prod_{n=1}^{\infty} \left(1 - \frac{s}{\rho_n}\right) e^{\frac{s}{\rho_n}}. \tag{5}$$

Corollary 2. *The nontrivial zeros of the zeta function are distributed symmetrically with respect to the lines $\mathrm{Re}\ s = 1/2$ and $\mathrm{Im}\ s = 0$.*

We shall henceforth enumerate the nontrivial zeros of the zeta function in order of the increasing absolute value of their imaginary parts, where zeros whose imaginary parts have the same absolute value are arranged arbitrarily.

Theorem 3. *We have the formula*

$$\frac{\zeta'(s)}{\zeta(s)} = -\frac{1}{s-1} + \sum_{n=1}^{\infty} \left(\frac{1}{s-\rho_n} + \frac{1}{\rho_n}\right) + \sum_{n=1}^{\infty} \left(\frac{1}{s+2n} - \frac{1}{2n}\right) + B_0,$$

where the ρ_n are the nontrivial zeros of $\zeta(s)$ and B_0 is an absolute constant.

Proof. The result follows from logarithmic differentiation of the left and right sides of (5). $\qquad \square$

Theorem 4. *Let $\rho_n = \beta_n + i\gamma_n$, $n = 1, 2, 3, \ldots$ be the nontrivial zeros of $\zeta(s)$, and let $T \geq 2$. Then*

$$\sum_{n=1}^{\infty} \frac{1}{1 + (T - \gamma_n)^2} \leq c \log T.$$

Proof. Let $s = 2 + iT$. Then

$$\left| \sum_{n=1}^{\infty} \left(\frac{1}{s + 2n} - \frac{1}{2n} \right) \right| \leq \sum_{n \leq T} \left(\frac{1}{2n} + \frac{1}{2n} \right) + \sum_{n > T} \frac{|s|}{4n^2} \leq c_0 \log T, \qquad (6)$$

By Theorem 3,

$$- \operatorname{Re} \frac{\zeta'(s)}{\zeta(s)} = \operatorname{Re} \left(\frac{1}{s - 1} - B_0 - \sum_{n=1}^{\infty} \left(\frac{1}{s + 2n} - \frac{1}{2n} \right) \right)$$

$$- \operatorname{Re} \sum_{n=1}^{\infty} \left(\frac{1}{s - \rho_n} + \frac{1}{\rho_n} \right) \leq c_1 \log T - \operatorname{Re} \sum_{n=1}^{\infty} \left(\frac{1}{s - \rho_n} + \frac{1}{\rho_n} \right).$$

Since

$$\left| \frac{\zeta'(s)}{\zeta(s)} \right| = \left| \sum_{n=1}^{\infty} \frac{\Lambda(n)}{n^{2 + iT}} \right| < c_2,$$

it follows that

$$\operatorname{Re} \sum_{n=1}^{\infty} \left(\frac{1}{s - \rho_n} + \frac{1}{\rho_n} \right) \leq c_3 \log T.$$

The Theorem now follows from the inequalities

$$\operatorname{Re} \frac{1}{s - \rho_n} = \operatorname{Re} \frac{1}{(2 - \beta_n) + i(T - \gamma_n)}$$

$$= \frac{2 - \beta_n}{(2 - \beta_n)^2 + (T - \gamma_n)^2} \geq \frac{0,5}{1 + (T - \gamma_n)^2},$$

$$\operatorname{Re} \frac{1}{\rho_n} = \frac{\beta_n}{\beta_n^2 + \gamma_n^2} \geq 0, \qquad \qquad \square$$

Corollary 1. *The number of zeros ρ_n of the zeta function that satisfy $T \leq |\operatorname{Im} \rho_n| \leq T + 1$ does not exceed $c_4 \log T$.*

Corollary 2. *For $T \geq 2$, we have*

$$\sum_{|T - \gamma_n| > 1} \frac{1}{|T - \gamma_n|^2} = O(\log T).$$

Corollary 3. *For $s = \sigma + it$, where $-1 \leq \sigma \leq 2$ and $|t| \geq 2$, we have*

$$\frac{\zeta'(s)}{\zeta(s)} = \sum_{|t - \gamma_n| \leq 1} \frac{1}{s - \rho_n} + O(\log |t|),$$

where the summation runs over all zeros ρ_n of the function $\zeta(s)$ that satisfy $|t - \operatorname{Im} \rho_n| \leq 1$.

Proof. Since the estimate (6) holds for all $s = \sigma + it$ with $|t| \geq 2$ and $-1 \leq \sigma \leq 2$, it follows that

$$\frac{\zeta'(s)}{\zeta(s)} = \sum_{n=1}^{\infty} \left(\frac{1}{s - \rho_n} + \frac{1}{\rho_n} \right) + O(\log|t|).$$

From this equation we subtract the same equation with $s = 2 + it$, and obtain

$$\frac{\zeta'(s)}{\zeta(s)} = \sum_{n=1}^{\infty} \left(\frac{1}{s - \rho_n} - \frac{1}{2 + it - \rho_n} \right) + O(\log|t|).$$

If $|\gamma_n - t| > 1$, then

$$\left| \frac{1}{\sigma + it - \rho_n} - \frac{1}{2 + it - \rho_n} \right| \leq \frac{2 - \sigma}{(\gamma_n - t)^2} \leq \frac{3}{(\gamma_n - t)^2}$$

and the result follows from Corollaries 1 and 2.   □

Theorem 5 (de la Vallée–Poussin). *There exists an absolute constant $c > 0$ such that there are no zeros of the zeta function in the domain*

$$\mathrm{Re}\, s = \sigma \geq 1 - \frac{c}{\log(|t| + 2)}.$$

Proof. At the point $s = 1$ the function $\zeta(s)$ has a pole, and so there exists a positive constant γ_0 such that the zeta function has no zeros in the domain $|s - 1| \leq \gamma_0$. Let $\rho_n = \beta_n + i\gamma_n$ be a zero of $\zeta(s)$ such that $|\gamma_n| > \gamma_0$. For $\mathrm{Re}\, s = \sigma > 1$

$$-\frac{\zeta'(s)}{\zeta(s)} = \sum_{n=1}^{\infty} \frac{\Lambda(n)}{n^s} = \sum_{n=1}^{\infty} \Lambda(n) n^{-\sigma} e^{-it \log n},$$

and so

$$-\mathrm{Re}\, \frac{\zeta'(s)}{\zeta(s)} = \sum_{n=1}^{\infty} \Lambda(n) n^{-\sigma} \cos(t \log n).$$

Since the inequality

$$3 + 4\cos\varphi + \cos^2\varphi = 2(1 + \cos\varphi)^2 \geq 0,$$

holds for every real number φ, it follows that

$$3 \left\{ -\frac{\zeta'(\sigma)}{\zeta(\sigma)} \right\} + 4 \left\{ -\mathrm{Re}\, \frac{\zeta'(\sigma + it)}{\zeta(\sigma + it)} \right\} + \left\{ -\mathrm{Re}\, \frac{\zeta'(\sigma + i2t)}{\zeta(\sigma + i2t)} \right\} \geq 0. \qquad (7)$$

We shall find upper estimates for each summand on the left side of (7). From Theorem 3 and Corollary 1 of Theorem 4, we obtain for $s = \sigma$

$$-\frac{\zeta'(\sigma)}{\zeta(\sigma)} < \frac{1}{\sigma - 1} + B_1,$$

where $B_1 > 0$ is an absolute constant. Further, from Theorem 3 for $s = \sigma + it$, where $1 < \sigma \le 2$ and $|t| > \gamma_0$, we find that

$$- \operatorname{Re} \frac{\zeta'(s)}{\zeta(s)} < A \log(|t| + 2) - \sum_{k=1}^{\infty} \operatorname{Re}\left(\frac{1}{s - \rho_k} + \frac{1}{\rho_k}\right),$$

where $A > 0$ is an absolute constant. Since $0 \le \beta_k \le 1$, $\rho_k = \beta_k + i\gamma_k$,

$$\operatorname{Re} \frac{1}{s - \rho_k} = \operatorname{Re} \frac{1}{\sigma - \beta_k + i(t - \gamma_k)} = \frac{\sigma - \beta_k}{(\sigma - \beta_k)^2 + (t - \gamma_k)^2} > 0.$$

Moreover,

$$\operatorname{Re} \frac{1}{\rho_k} = \frac{\beta_k}{\beta_k^2 + \gamma_k^2} \ge 0.$$

Therefore,

$$- \operatorname{Re} \frac{\zeta'(\sigma + it)}{\zeta(\sigma + it)} < A \log(|t| + 2) - \frac{\sigma - \beta_n}{(\sigma - \beta_n)^2 + (t - \gamma_n)^2}.$$

It follows from this inequality that

$$- \operatorname{Re} \frac{\zeta'(\sigma + i2t)}{\zeta(\sigma + i2t)} < A \log(2|t| + 2).$$

Putting these estimates into (7), we obtain

$$\frac{3}{\sigma - 1} - 4\frac{\sigma - \beta_n}{(\sigma - \beta_n)^2 + (t - \gamma_n)^2} + A_1 \log(|t| + 2) \ge 0,$$

where A_1 is an absolute constant. The last inequality holds for any t such that $|t| > \gamma_0$ and for any σ such that $1 < \sigma \le 2$. Let us apply the inequality with $t = \gamma_n$ and $\sigma = 1 + \dfrac{1}{2A_1 \log(|\gamma_n| + 2)}$. We find that

$$\frac{4}{\sigma - \beta_n} \le \frac{3}{\sigma - 1} + A_1 \log(|\gamma_n| + 2),$$

$$\beta_n \le 1 - \frac{1}{14A_1 \log(|\gamma_n| + 2)},$$

which is what we had to prove.   □

Corollary. *Let* $T \ge 2$ *and let* $c > 0$ *be the absolute constant in the Theorem. Then in the domain*

$$\sigma \ge 1 - \frac{c}{2\log(T + 2)}, \quad 2 \le |t| \le T,$$

we have the estimate

$$\left|\frac{\zeta'(s)}{\zeta(s)}\right| = 0(\log^2 T).$$

Proof. From Corollary 3 of Theorem 4, we have

$$\frac{\zeta'(s)}{\zeta(s)} = \sum_{|t - \gamma_n| \le 1} \frac{1}{s - \rho_n} + O(\log T),$$

and so

$$\left|\frac{\zeta'(s)}{\zeta(s)}\right| \le \sum_{|t - \gamma_n| \le 1} \frac{1}{|\sigma - \beta_n + i(t - \gamma_n)|} + O(\log T).$$

Since

$$\beta_n \le 1 - \frac{c}{\log(T + 2)}, \quad \sigma \ge 1 - \frac{c}{2\log(T + 2)},$$

then

$$\left|\frac{\zeta'(s)}{\zeta(s)}\right| \le \frac{2}{c} \log(T + 2) \sum_{|t - \gamma_n| \le 1} 1 + O(\log T) = O(\log^2 T),$$

which is what we had to prove.   □

§3. Approximation by a Finite Sum

In order to solve many problems in number theory, we need to estimate $|\zeta(s)|$ in the critical strip. Since $\zeta(s)$ can be approximated by the sum of the first terms of the series that defines it for Re $s > 1$, we shall estimate this sum. We obtain below the simplest approximation to $\zeta(s)$.

Theorem 6. *For $0 < \sigma_0 \le \sigma \le 2$ and $2\pi \le |t| \le \pi x$, we have*

$$\zeta(s) = \sum_{n \le x} \frac{1}{n^s} + \frac{x^{1-s}}{s - 1} + O(x^{-\sigma} \ln x),$$

where the constant in the O symbol depends only on σ_0.

Proof. By Lemma 2, for $N > x$,

$$\zeta(s) = \sum_{n=1}^{N} \frac{1}{n^s} - \frac{N^{1-s}}{1 - s} - \frac{1}{2} N^{-s} + s \int_{N}^{\infty} \frac{\frac{1}{2} - \{u\}}{u^{s+1}} \, du. \tag{8}$$

The last summand has order $O(|t| N^{-\sigma})$. We consider the sums S,

$$S = \sum_{x < n \le N} \frac{1}{n^s}, \quad s = \sigma + it.$$

Introducing the notation $\phi(n) = n^{-\sigma}$, $f(n) = -(t/2\pi) \ln n$, we can apply Lemma 1 of Chapter I to S and obtain

$$S = \int_{x}^{N} u^{-s} \, du + O(x^{-\sigma} \ln x) = \frac{N^{1-s}}{1 - s} + \frac{x^{1-s}}{s - 1} + O(x^{-\sigma} \ln x).$$

From this and from (8), and letting $N \to \infty$, we obtain the assertion of the Theorem.

Exercises

1. Prove that if $|\arg| < \pi/2$, then

$$\Gamma\left(\frac{s}{2}\right)\zeta(s) = \sum_{n=1}^{\infty} \frac{\Gamma\left(\frac{s}{2}, \pi\tau n^2\right)}{n^s}$$

$$+ \pi^{s-\frac{1}{2}} \sum_{n=1}^{\infty} \frac{\Gamma\left(\frac{1-s}{2}, \frac{\pi n^2}{\tau}\right)}{n^{1-s}} - \pi^{\frac{1}{2}} \frac{(\pi\tau)^{\frac{s-1}{2}}}{1-s} - \frac{(\pi\tau)^{\frac{s}{2}}}{1-s},$$

where $\Gamma(z, x)$ is the incomplete gamma function, defined by

$$\Gamma(z, x) = \int_x^{\infty} e^{\xi} \xi^{z-1} d\xi.$$

2. Deduce the following from the formula in exercise 1:

a) $\pi^{-\frac{s}{2}}\Gamma\left(\frac{s}{2}\right)\zeta(s) = \pi^{-\frac{1-s}{2}}\Gamma\left(\frac{1-s}{2}\right)\zeta(1-s)$;

b) Let $h > 0$, $2\pi xy = |t|$, $x > h$, $y > h$, $K > 0$. Then for $-K < \sigma < K$,

$$\zeta(s) = \sum_{n \le x} \frac{1}{n^s} + \pi^{s-\frac{1}{2}} \frac{\Gamma\left(\frac{1-s}{2}\right)}{\Gamma\left(\frac{s}{2}\right)} \sum_{n \le y} \frac{1}{n^{1-s}} + O(x^{-\sigma}) + O\left(|t|^{\frac{1}{2}-\sigma} y^{\sigma-1}\right).$$

3. Show that the function $Z(t)$ defined by

$$Z(t) = e^{i\theta(t)} \zeta\left(\frac{1}{2} + it\right),$$

where

$$e^{i\theta(t)} = \frac{\pi^{-it/2}\Gamma\left(\frac{1}{4} + \frac{it}{2}\right)}{\left|\Gamma\left(\frac{1}{4} + \frac{it}{2}\right)\right|},$$

is real-valued for real values of t.

4. Prove that the function $\theta(t)$, $t \geq 2$, defined in exercise 3, has the form

$$\theta(t) = t \ln \sqrt{\frac{t}{2\pi}} - \frac{t}{2} - \frac{\pi}{8} + \Delta(t),$$

$$\Delta(t) = \frac{t}{4} \ln \left(1 + \frac{1}{4t^2}\right) + \frac{1}{4} \operatorname{arctg} \frac{1}{2t} - \frac{t}{2} \int_0^\infty \frac{\rho(u)\, du}{\left(u + \frac{1}{4}\right)^2 + \frac{t^2}{4}},$$

$$\rho(u) = \frac{1}{2} - \{u\}.$$

5. For any integer $k \geq 0$ and $t \geq 2$, show that

$$Z^{(2k)}(t) = (-1)^k \cdot 2 \sum_{n \leq \sqrt{t/2\pi}} \frac{(\theta'(t) - \ln n)^{2k}}{\sqrt{n}} \cos(\theta(t) - t \ln n)$$

$$+ O(t^{-\frac{1}{4}(\ln t)^{2k+1}}),$$

$$Z^{(2k+1)}(t) = (-1)^{k+1} 2 \sum_{n \leq \sqrt{t/2\pi}} \frac{(\theta'(t) - \ln n)^{2k+1}}{\sqrt{n}} \sin(\theta(t) - t \ln n)$$

$$+ O(t^{-\frac{1}{4}(\ln t)^{2k+2}}),$$

where the constant in the symbol O depends only on k.

6. Let $k \geq 0$ be a fixed integer, let $T \geq 2$ and $P = \sqrt{T/2\pi}$, and define the function $\Phi(t)$ by

$$\Phi(t) = \sum_{n \leq P} \frac{\left(\ln \frac{P}{n}\right)^k}{\sqrt{n}} \cos\left(t \ln \frac{P}{n}\right) + O\left(T^{-\frac{1}{4}}(\ln T)^{k+1}\right).$$

Further, let

$$v_0 = \left[\frac{T \ln P}{\pi}\right] + 1, \qquad r = [\ln T], \qquad H_1 = \left[\frac{H \ln P}{\pi r}\right],$$

$$v = v_0 + v_1 + \ldots + v_r, \qquad 0 \leq v_1, \ldots, v_r \leq H_1 - 1.$$

Equating $|S_1|$ and $|S_2|$, where

$$S_1 = \sum_{v_1 = 0}^{H_1} \ldots \sum_{v_r = 0}^{H_1 - 1} \Phi(t_v), \qquad S_2 = \sum_{v_1 = 0}^{H_1 - 1} \ldots \sum_{v_r = 0}^{H_1 - 1} (-1)^v \Phi(t_v),$$

prove the existence of a zero of order of the function $\Phi(t)$ on the interval $(T, T + H)$, where

$$H \geq T^{\frac{1}{6k+6}}(\ln T)^2, \qquad T \geq T_0.$$

7. Let $k \geq 0$ be an integer, and let $T \geq T_0$,

$$H \geq T^{\frac{1}{6k+6}}(\ln T)^2.$$

Show that the interval $(T, T + H)$ contains a zero of order of the function $Z^{(k)}(t)$.

Chapter V. The Connection Between the Sum of the Coefficients of a Dirichlet Series and the Function Defined by this Series

The method used to prove the theorems in this chapter is called the method of complex integration.

§1. A General Theorem

Definition 1. A *Dirichlet series* is an expression of the form

$$f(s) = \sum_{n=1}^{\infty} \frac{a_n}{n^s}, \tag{1}$$

where the complex numbers a_n are the coefficients of the Dirichlet series, and $s = \sigma + it$.

We shall study the sum function $\Phi(x) = \sum_{n \leq x} a_n$ of the coefficients of the Dirichlet series. The function $\Phi(x)$ can be expressed in terms of the series $f(s)$.

Theorem 1. *Assume that the series (1) for $f(s)$ converges absolutely for $\sigma > 1$, that $|a_n| \leq A(n)$, where $A_n > 0$ is a monotonically increasing function, and that*

$$\sum_{n=1}^{\infty} |a_n| n^{-\sigma} = O((\sigma - 1)^{-\alpha}), \quad \alpha > 0.$$

as $\sigma \to 1 + 0$. Then for any $b_0 \geq b > 1$, $T \geq 1$, and $x = N + 1/2$, we have

$$\Phi(x) = \sum_{n \leq x} a_n = \frac{1}{2\pi i} \int_{b-iT}^{b+iT} f(s) \frac{x^s}{s} \, ds + O\left(\frac{x^b}{T(b-1)^\alpha}\right) + O\left(\frac{xA(2x)\log x}{T}\right),$$

where the constants in the O symbols depend only on b_0.

Proof. First, we note that

$$\frac{1}{2\pi i} \int_{b-iT}^{b+iT} \frac{a^s}{s} \, ds = \begin{cases} 1 + O\left(\dfrac{a^b}{T|\log a|}\right), & \text{if } a > 1; \\[2mm] O\left(\dfrac{a^b}{T|\log a|}\right), & \text{if } 0 < a < 1. \end{cases} \tag{2}$$

Indeed, let $a > 1$ (resp. $0 < a < 1$). Choose $U > b$, and consider the contour Γ (resp. Γ_1), sketched in Fig. 5.

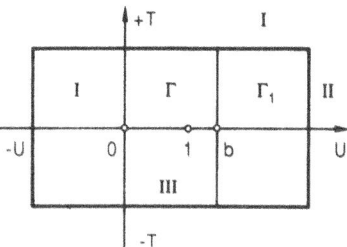

Fig. 5

By Cauchy's theorem, for $a > 1$,

$$\frac{1}{2\pi i} \int_{\Gamma} \frac{a^s \, ds}{s} = 1$$

and, for $a < 1$,

$$\frac{1}{2\pi i} \int_{\Gamma_1} \frac{a^s \, ds}{s} = 0$$

i.e.

$$\frac{1}{2\pi i} \int_{b-iT}^{b+iT} \frac{a^s \, ds}{s} = 1 + R \qquad (a > 1),$$

$$\frac{1}{2\pi i} \int_{b-iT}^{b+iT} \frac{a^s \, ds}{s} = 1 + R_1 \qquad (0 < a < 1),$$

(3)

where R and R_1 are the corresponding integrals along the sides I, II, III. The integrals along I and III are equal in absolute value. Therefore, if $a > 1$, then

$$\frac{1}{2\pi i} \left| \int_I \frac{a^s}{s} \, ds \right| \le \frac{1}{2\pi} \int_{-U}^{b} \frac{a^\sigma \, d\sigma}{\sqrt{T^2 + \sigma^2}} \le \frac{a^b}{T \log a}$$

and, if $0 < a < 1$, then

$$\frac{1}{2\pi} \left| \oint_I \frac{a^s}{s} \, ds \right| \le \frac{1}{2\pi} \int_b^U \frac{a^\sigma \, d\sigma}{\sqrt{T^2 + \sigma^2}} \le \frac{a^b}{T |\log a|}.$$

Moreover, if $a > 1$, then

$$\frac{1}{2\pi i} \left| \int_{II} \frac{a^s}{s} \, ds \right| \le \frac{1}{2\pi} \int_{-T}^{+T} \frac{a^{-U} \, dt}{\sqrt{U^2 + t^2}} = O(a^{-U}) \to 0$$

as $U \to +\infty$, and, if $0 < a < 1$, then

$$\frac{1}{2\pi} \left| \int_{II} \, ds \right| \le \frac{1}{2\pi} \int_{-T}^{+T} \frac{a^U \, dt}{\sqrt{U^2 + t^2}} = O(a^U) \to 0$$

as $U \to +\infty$. We obtain (2) by passing to the limit in (3) as $U \to \infty$. Since $x = N + 1/2$, then $x/n \neq 1$ for any natural number n. The series (1) converges absolutely for $s = b + it$. Integrating term by term, we find that

$$\frac{1}{2\pi i} \int_{b-iT}^{b+iT} f(s) \frac{x^s}{s} ds = \sum_{n=1}^{\infty} a_n \left(\frac{1}{2\pi i} \int_{b-iT}^{b+iT} \left(\frac{x}{n} \right)^2 \frac{ds}{s} \right) = \sum_{n \le x} a_n + R,$$

where

$$R = O\left(\sum_{n=1}^{\infty} |a_n| \left(\frac{x}{n} \right)^b T^{-1} \left| \log \frac{x}{n} \right|^{-1} \right).$$

We divide the sum in the O symbol into two parts. The first contains the summands for which $x/n \le 1/2$ or $x/n \ge 2$. For these terms,

$$\left| \log \frac{x}{n} \right| \ge \log 2.$$

Since, under the conditions of the Theorem,

$$\sum_{n=1}^{\infty} \frac{|a_n|}{n^b} = O\left(\frac{1}{(b-1)^\alpha} \right),$$

the first sum will be

$$O\left(\frac{x^b}{T(b-1)^\alpha} \right).$$

The second sum has the form

$$\sum_{\frac{1}{2}x < n < 2x} |a_n| \left(\frac{x}{n} \right)^b T^{-1} \left| \log \frac{x}{n} \right|^{-1} \le T^{-1} A(2x) 2^b \sum_{\frac{1}{2}x < n < 2x} \left| \log \frac{N+0,5}{n} \right|^{-1}.$$

Extracting from the last sum the terms with $n = N - 1$, N, and $N + 1$, whose values have order $O(x)$, we find that the remaining part of the sum satisfies the estimate

$$x \le \int_{x/2}^{N-1} \left(\log \frac{N+0,5}{u} \right)^{-1} du + \int_{N+1}^{2x} \left(\log \frac{u}{N+0,5} \right)^{-1} du = O(x \log x).$$

The Theorem follows from these estimates.   □

§2. The Prime Number Theorem

The asymptotic law for the distribution of the prime numbers, usually called the prime number theorem, is the statement that $\pi(x) \sim x/\ln x$, or equivalently, $\psi(x) \sim x$. This was first proved by J. Hadamard and C. de la Vallée Poussin. We shall now prove a stronger result.

Theorem 2 (de la Valleé Poussin). *There exists an absolute constant $c > 0$ such that*

$$\psi(x) = \sum_{n \leq x} \Lambda(n) = x + O(xe^{-c\sqrt{\ln x}});$$

$$\pi(x) = \sum_{p \leq x} 1 = \int_2^x \frac{du}{\ln u} + O\left(xe^{-\frac{c}{2}\sqrt{\ln x}}\right).$$

Proof. For Re $s > 1$, we have

$$-\frac{\zeta'(s)}{\zeta(s)} = \sum_{n=1}^{\infty} \frac{\Lambda(n)}{n^s}.$$

Without loss of generality, we can assume that $x = (N + 1/2) \geq 100$. We shall apply Theorem 1. It follows from the results in Chapter IV that in Theorem 1 we can take $\alpha = 1$ and $A(n) = \log n$. We set

$$b = 1 + \frac{1}{\log x}, \quad T = e^{\sqrt{\log x}}.$$

Then

$$\psi(x) = \sum_{n \leq x} \Lambda(n) = \frac{1}{2\pi i} \int_{b-iT}^{b+iT} \left(-\frac{\zeta'(s)}{\zeta(s)}\right) \frac{x^s}{s} \, ds + O\left(\frac{x \ln^2 x}{T}\right).$$

By Theorem 5 of Chapter IV, and its Corollaries, there exists an absolute constant $c_1 > 0$ such that the zeta function has no zeros in the domain Re $s = \sigma \geq \sigma_1 = 1 - \dfrac{c_1}{2\log(T+2)}$ and $|t| \leq T$. Moreover,

$$\frac{\zeta'(s)}{\zeta(s)} = O(\log^2 T)$$

for $s = \sigma_i + it$, $s = \sigma \pm iT$, $\sigma_1 \leq \sigma \leq b$, $s = b + it$.

We shall consider the integral J around the contour Γ (cf. Fig. 6), where J is defined by

$$J = \frac{1}{2\pi i} \int_{\Gamma} \left(-\frac{\zeta'(s)}{\zeta(s)}\right) \frac{x^s}{s} \, ds.$$

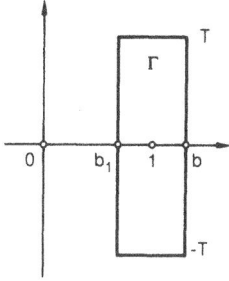

Fig. 6

Inside the contour, the integrand has a pole of order one with residue equal to x. Therefore,

$$\frac{1}{2\pi i}\int_{b-iT}^{b+iT}\left(-\frac{\zeta'(s)}{\zeta(s)}\right)\frac{x^s}{s}\,ds = x + R,$$

where R is the sum of the integrals along the top, bottom, and left sides of Γ. We shall estimate these integrals. The first two are equal in absolute value and can be estimated as follows:

$$\left|\frac{1}{2\pi i}\int_{\sigma_1+iT}^{b+iT}\left(-\frac{\zeta'(s)}{\zeta(s)}\right)\frac{x^s}{s}\,ds\right| \le \int_{\sigma_1}^{b}\left|\frac{\zeta'(\sigma+iT)}{\zeta(\sigma+iT)}\right|\frac{x^\sigma}{T}\,d\sigma = O\left(\frac{x\log^2 T}{T}\right)$$

The integral on the left side equals

$$\left|\frac{1}{2\pi i}\int_{\sigma_1-iT}^{\sigma_1+iT}\left(-\frac{\zeta'(s)}{\zeta(s)}\right)\frac{x^s}{s}\,ds\right| = \left|\frac{1}{2\pi i}\int_{-T}^{+T}\frac{\zeta'(\sigma_1+it)x^{\sigma_1+it}}{\zeta(\sigma_1+it)(\sigma_1+it)}\,dt\right|$$

$$= O\left(x^{\sigma_1}\log^2 T\left(\int_0^1\frac{dt}{\sigma_1}+\int_1^T\frac{dt}{t}\right)\right) = O(x^{\sigma_1}\log^3 T).$$

The first assertion of the Theorem now follows from these estimates and the definitions of T and σ_1.

Now consider

$$S = \sum_{n\le x}\frac{\Lambda(n)}{\log n} = \sum_{p\le x}1 + \sum_{\substack{x\le n=p^k\\ k\ge 2}}\frac{\Lambda(n)}{\log n}.$$

In the second sum we have $k \le \log x$, and for each $k \ge 2$ the sum contains $\le x$ summands, each no greater than one. Therefore,

$$S = \pi(x) + O(\sqrt{x}\log x). \tag{4}$$

Now apply Lemma C of Chapter I with $c_n = \Lambda(n)$ and $f(x) = 1/\log x$. Then

$$C(x) = \sum_{n\le x}c_n = \psi(x) = x + O(xe^{-c\sqrt{\log x}}),\quad f'(x) = -1/x\log^2 x.$$

This gives

$$S = \int_2^x\frac{\psi(u)}{u\log^2 u}\,du + \frac{\psi(x)}{\log x} = \int_2^x\frac{du}{\log^2 u} + \frac{x}{\log x} + R,$$

where

$$R = O\left(\int_2^x e^{-c\sqrt{\log u}}\frac{du}{\log^2 u} + xe^{-c\sqrt{\log x}}\right)$$

$$= O\left(\int_2^{\sqrt{x}}x\,du + \int_{\sqrt{x}}^x c\sqrt{\log u}\,du + xe^{-\sqrt{\log x}}\right) = O\left(xe^{-\frac{c}{2}\sqrt{\ln x}}\right)$$

and

$$\int_2^x\frac{du}{\log^2 u} + \frac{x}{\log x} = -\left.\frac{u}{\log u}\right|_2^x + \int_2^x\frac{du}{\log u} + \frac{x}{\log x} = \int_2^x\frac{du}{\log u} + \frac{2}{\log 2}.$$

The second assertion of the Theorem now follows from this identity and from (4). This completes the proof of the Prime Number Theorem.   □

§3. Representation of the Chebyshev Functions as Sums Over the Zeros of the Zeta Function

The method of complex integration allows us to write down explicit formulae connecting various kinds of sums over primes with the zeros of the zeta function. We shall now prove one such formula.

Theorem 3. *Let* $2 \le T \le x$. *Then*

$$\psi(x) = \sum_{n \le x} \Lambda(n) = x - \sum_{|\operatorname{Im}\rho| \le T} \frac{x^\rho}{\rho} + O\left(\frac{x \log^2 x}{T}\right),$$

where the ρ are the zeros of the zeta function in the critical strip.

Proof. By Theorem 3 of Chapter IV for Re $s > 1$

$$-\frac{\zeta'(s)}{\zeta(s)} = \sum_{n=1}^{\infty} \frac{\Lambda(n)}{n^s}$$

$$= \frac{1}{s-1} - \sum_{n=1}^{\infty}\left(\frac{1}{s-\rho_n}+\frac{1}{\rho_n}\right) - \sum_{n=1}^{\infty}\left(\frac{1}{s+2n}-\frac{1}{2n}\right) - B_0, \tag{5}$$

where the ρ_n are the nontrivial zeros of $\zeta(s)$. Just as in the proof of Theorem 2, with $b = 1 + 1/\log x$,

$$\psi(x) = \frac{1}{2\pi i}\int_{b-iT_1}^{b+iT_1}\left(-\frac{\zeta'(s)}{\zeta(s)}\right)\frac{x^s}{s}\,ds + O\left(\frac{x\log^2 x}{T}\right), \tag{6}$$

where $T \le T_1 \le T+1$ and T_1 is chosen so that the distance from the line Im $s = T_1$ to the nearest zero of $\rho(s)$ is $\gg 1/\log T$. This is always possible, since, by Corollary 1 to Theorem 4 of Chapter IV, the number of zeros of $\rho(s)$ for which $T \le \operatorname{Im}\rho \le T+1$ is $O(\log T)$. Consider the integral J, defined by

$$J = \frac{1}{2\pi i}\int_\Gamma \left(-\frac{\zeta'(s)}{\zeta(s)}\right)\frac{x^s}{s}\,ds,$$

where Γ is the rectangle shown in Fig. 7. By Cauchy's theorem and (5)

$$J = x - \sum_{|\operatorname{Im}\rho| \le T_1}\frac{x^\rho}{\rho} - \frac{\zeta'(0)}{\zeta(0)}. \tag{7}$$

It remains to estimate the integrals over the sides I, II, and III of Γ. The integrals over sides I and III are equal in absolute value, and can be estimated from above

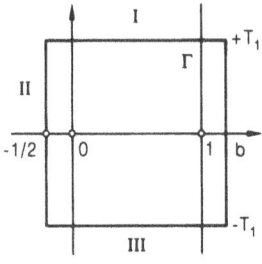

Fig. 7

by

$$\frac{1}{2\pi}\left|\int_{-0.5+iT_1}^{b+iT_1}\left(-\frac{\zeta'(s)}{\zeta(s)}\right)\frac{x^s}{s}\,ds\right| < \frac{x}{T_1}\int_{-0.5}^{b}\left|\frac{\zeta'(\sigma+iT_1)}{\zeta(\sigma+iT_1)}\right|d\sigma. \qquad (8)$$

The integral over II does not exceed

$$\frac{1}{2\pi}\left|\int_{-0.5-iT_1}^{-0.5+iT_1}\left(-\frac{\zeta'(s)}{\zeta(s)}\right)\frac{x^s}{s}\,ds\right| < \frac{1}{\sqrt{x}}\int_{-T_1}^{T_1}\left|\frac{\zeta'(-0.5+it)}{\zeta(-0.5+it)}\right|\frac{dt}{(\log x)^{-1}+|t|}. \qquad (9)$$

We estimate the value of $\left|\dfrac{\zeta'(\sigma+it)}{\zeta(\sigma+it)}\right|$, where either $-1/2 \le \sigma \le b$ and $t = T_1$, or $\sigma = -1/2$ and $2 \le |t| \le T_1$. Again, by Corollary 3 to Theorem 4 of Chapter IV,

$$\left|\frac{\zeta'(\sigma+it)}{\zeta(\sigma+it)}\right| = \sum_{|t-\gamma_n|\le 1}\frac{1}{\sigma-\sigma_n+i(t-\gamma_n)} + O(\log(|t|+2)).$$

The last sum has order $O(\log^2 x)$, since, if $|t| \le T_1$, then $\sigma = -1/2$ and the number of zeros of $\rho(s)$ such that $|t - \gamma_n| \le 1$ does not exceed $O(\log(|t|+2))$. If $t = T_1$ and $-1/2 \le \sigma \le b$, then it follows from the choice of T_1 that

$$|T_i - \gamma_n| \gg (\log T)^{-1}.$$

The Theorem now follows from the estimates (8), (9), and (7).   □

Remark. If we take $T = \exp(\sqrt{\ln x})$ in Theorem 3, then Theorems 4 and 5 of Chapter IV give another proof of Theorem 2.

Exercises

1. Suppose that the Dirichlet series $f(s) = \sum_{n=1}^{\infty}\dfrac{a_n}{n^s}$ converges absolutely for

Re $s > 1$. For some $b > 1$, let $A(\xi) = \sum_{n\le\xi} a_n$, and $B(b) = \int_1^{\infty}\dfrac{|A(\xi)|}{\xi^{b+1}}\,d\xi$. Then for

$x \geq 1$ and $T \geq 2$ we have

$$\int_1^x A(\xi) \, d\xi = \frac{1}{2\pi i} \int_{b-iT}^{b+iT} \frac{f(s)}{s(s+1)} x^{s+1} \, ds + R,$$

where

$$|R| \leq c \left\{ B(b) \frac{x^{b+1}}{T} + 2^b \left(\frac{x \log x}{T} + \log T \right) \max_{\frac{1}{2}x \leq \xi \leq \frac{3}{2}x} |A(\xi)| \right\}.$$

2. Prove that for $T \geq 2$

$$\int_T^{2T} \left| \zeta \left(\frac{1}{2} + it \right) \right|^4 \, dt = O(T \ln^4 T).$$

3. Prove that for $X \geq 2$

$$\sum_{n \leq X} \tau_4(n) = X P_3(\ln X) + O(\sqrt{X} \ln^5 X),$$

where $P_3(u)$ is polynomial of degree three.

4. Prove that for $X \geq 2$

$$M(X) = \sum_{n \leq X} \mu(n) = O(X e^{-c\sqrt{\log X}}).$$

5. Prove the following: In order that $\rho(s)$ have no zeros in the half plane $\operatorname{Re} s > \gamma$, where γ satisfies $1/2 < \gamma < 1$, it is necessary and sufficient that one of the following conditions be satisfied for arbitrarily small ε:

a) $\psi(x) = x + O'(x^{\gamma+\varepsilon})$;

b) $\pi(x) = \int_2^x \frac{du}{\log u} + O(x^{\gamma+\varepsilon})$;

c) $M(x) = O(x^{\gamma+\varepsilon})$.

6. Let $0 \leq \alpha < 1$ and $N \geq 2$. Then

$$S = \sum_{n=2}^{\infty} e^{-n/N} e^{-2\pi i \alpha n} \Lambda(n) = \frac{1}{x} - \sum_{\rho_k} x^{-\rho_k} \Gamma(\rho_k) + O(\log^3 N),$$

where the ρ_k are the zeros of $\zeta(s)$ and $x = 1/N + 2\pi i \alpha$.

7. Define $\psi_0(x) = \dfrac{\psi(x+0) + \psi(x-0)}{2}$. Prove that for $x \geq 2$

$$\psi_0(x) = x - \sum_{n=1}^{\infty} \frac{x^{\rho_n}}{\rho_n} - \frac{\zeta'(0)}{\zeta(0)} - \frac{1}{2} \log \left(1 - \frac{1}{x^2} \right).$$

8. a) Let $r(n)$ denote the number of solutions of the equation $\varphi(m) = n$. Then

$$\sum_{n \leq x} r(n) = c_0 x + O \left(\frac{x}{\ln x} \right).$$

b) Prove that

$$\sum_{n \le x} \frac{1}{\varphi(n)} = c_0 \ln x + O(\ln \ln x).$$

9. Let $x > 2$. Then

$$\sum_{p \le x} \frac{1}{p} = c_1 + \ln \ln x + O\left(\frac{1}{\ln x}\right),$$

$$\prod_{p \le x} \left(1 - \frac{1}{p}\right) = \frac{c_2}{\ln x} + O\left(\frac{1}{\ln^2 x}\right).$$

Chapter VI. The Method of I.M. Vinogradov in the Theory of the Zeta Function

In this chapter we shall prove a mean value theorem due to I.M. Vinogradov, and from it deduce an estimate for the zeta function in a neighborhood of Re $s = 1$, a new boundary for zeros of the zeta function, and a new remainder term in the prime number theorem.

§1. Theorem on the Mean Value of the Modulus of a Trigonometric Sum

By the mean value of the modulus of a trigonometric sum we mean the value of the integral J, where

$$J = J_{k,n}(P) = \int_0^1 \ldots \int_0^1 \left| \sum_{x \leq P} e^{2\pi i(\alpha_1 x + \ldots + \alpha_n x^n)} \right|^{2k} d\alpha_1 \ldots d\alpha_n.$$

In many problems in number theory it is important to know the order of magnitude of the growth of J with increasing P (the mean value theorem).

We shall first prove several auxiliary results.

Lemma 1. Let $\lambda_1, \ldots, \lambda_n$ be integers, and let $J_{k,n}(\lambda_1, \ldots, \lambda_n)$ denote the number of solutons of the system of equations

$$\begin{cases} x_1 + \ldots - x_{2k} = \lambda_1, \\ \ldots\ldots\ldots\ldots\ldots \\ x_1^n + \ldots - x_{2k}^n = \lambda_n, \\ \quad 1 \leq x_1, \ldots, x_{2k} \leq P. \end{cases} \tag{1}$$

Then the following relations hold:

a)

$$J_{k,n}(\lambda_1, \ldots, \lambda_n) =$$

$$= \int_0^1 \ldots \int_0^1 \left| \sum_{x \leq P} e^{2\pi i(\alpha_1 x + \ldots + \alpha_n x^n)} \right|^{2k} e^{-2\pi i(\alpha_1 \lambda_1 + \ldots + \alpha_n \lambda_n)} d\alpha_1 \ldots d\alpha_n;$$

b)

$$J_{k,n}(\lambda_1, \dots, \lambda_n) \le J_{k,n}(0, \dots, 0) = J_{k,n}(P) = J;$$

c)

$$\sum_{\lambda_1, \dots, \lambda_n} J_{k,n}(\lambda_1, \dots, \lambda_n) = P^{2k};$$

d)

$$|\lambda_1| < kP, \dots, |\lambda_n| < kP^n;$$

e)

$$J = J_{k,n}(P) > (2k)^{-n} P^{2k - \frac{n^2+n}{2}};$$

f) If $x_1, \dots, x_{2k}$ satisfy (1) with $\lambda_1 = \dots = \lambda_n = 0$, then $x_1 + A, \dots, x_{2k} + A$ satisfy (1) with $\lambda_1 = \dots = \lambda_n = 0$ for any A.

Proof. For any integer λ we have

$$\int_0^1 e^{2\pi i \alpha \lambda} d\alpha = \begin{cases} 1, & \text{if } \lambda = 0; \\ 0, & \text{if } \lambda \ne 0. \end{cases}$$

This implies a), if we raise the absolute value of the integrand to the $2k$-th power and integrate over $\alpha_1, \dots, \alpha_n$. Statement b) follows from the fact the absolute value of the integral does not exceed the integral of the absolute value of the integrand. Statement c) follows from the fact that the left side of the equation is P^{2k}, which is the number of all possible choices of $x_1, \dots, x_{2k}$ in system (1). Statement d) follows from the conditions on $x_1, \dots, x_{2k}$.

Statement e) follows from c), b), and d). Statement f) can be seen to hold by successively inserting the numbers $x_1 + A, \dots, x_{2k} + A$ into the first, second, ..., and last equations. This proves the Lemma. □

Lemma 2 (Hölder's inequality). *Let* $u_v, v_v \ge 0$, $\alpha > 0$, $\beta > 0$, $\beta > 0$, $\alpha + \beta = 1$. *Then*

$$\sum_{v=1}^P u_v v_v \le \left(\sum_{v=1}^P u_v^{1/\alpha} \right)^{\alpha} \left(\sum_{v=1}^P v_v^{1/\beta} \right)^{\beta}.$$

Proof. For $x \ge 1$ we have

$$x^\alpha \le \alpha x + \beta,$$

since the function $x^\alpha - \alpha x - \beta$ is decreasing. This implies that

$$a^\alpha b^\beta \le \alpha a + \beta b. \tag{2}$$

for $a, b \in [0, 1]$. If we define a and b by

$$a = \frac{u_v^{1/\alpha}}{\sum\limits_{v=1}^P u_v^{1/\alpha}}, \quad b = \frac{v_v^{1/\beta}}{\sum\limits_{v=1}^P v_v^{1/\beta}},$$

we obtain the statement of the Lemma. □

Corollaries. 1. *Cauchy's inequality:*

$$\left(\sum_{v=1}^{P} u_v v_v \right)^2 \leq \left(\sum_{v=1}^{P} u_v^2 \right)\left(\sum_{v=1}^{P} v_v^2 \right);$$

2.

$$\left(\sum_{v=1}^{P} u_v v_v \right)^k \leq \left(\sum_{v=1}^{P} u_v \right)^{k-1} \sum_{v=1}^{P} u_v v_v^k;$$

3.

$$\left(\sum_{v=1}^{P} u_v \right)^k \leq P^{k-1} \sum_{v=1}^{P} u_v^k.$$

4. *Let* $u_1 \geq 0$ *for* $i = 1, \ldots, k$. *Then*

$$u_1 \ldots u_k \leq \frac{u_1^k + \ldots + u_k^k}{k}$$

(the geometric mean of a set of nonnegative integers does not exceed the arithmetic mean). For $k = 1$ *the inequality is obvious. Assuming that the inequality holds for* $k - 1$ *and applying* (2), *we obtain*

$$u_1 \ldots u_k = ((u_1 \ldots u_{k-1})^{k/(k-1)})^{(k-1)/k}(u_k^k)^{1/k}$$

$$= \frac{k-1}{k}(u_1 \ldots u_{k-1})^{\frac{k}{k-1}} + \frac{1}{k}u_k^k \leq \frac{u_1^k + \ldots + u_k^k}{k}.$$

Lemma 3. *Let* $n > 2$, $P > (2n)^{4n}$, $H = (2n)^4$, *and let* R *be the least integer satisfying the condition* $HR \geq P$. *Let* $v_1, \ldots, v_n$ *run through the integers in the intervals*

$$X_i < v_i \leq Y_i, \ldots, X_n < v_n \leq Y_n,$$

where for some ω *satisfying* $0 \leq \omega < P$ *we have*

$$-\omega < X_i, \quad X_i + R = Y_1, \quad Y_1 + R \leq X_2, \ldots,$$

$$X_n + R = Y_n, \quad Y_n \leq -\omega + P.$$

Then the number E_1 *of systems of values* $v_1, \ldots, v_n$ *such that the sums* $V_1 = v_1 + \ldots + v_n, \ldots, V_n = v_1^n + \ldots + v_n^n$ *lie, respectively, in some intervals with lengths*

$$1, \ldots, P^{n-1}, \tag{3}$$

satisfies the inequality

$$E_1 < e^{r(n)-1} H^{\frac{n(n-1)}{2}}, \quad r(n) = -\frac{n^2}{2}\ln n + \frac{3}{4}n^2 + \frac{3}{2}n.$$

If $v_1', \ldots, v_n'$ *run through the same values as* $v_1, \ldots, v_n$ *(independently of the latter), then the number* E *of cases for which the differences* $V_1 - V_1', \ldots,$

$V_n - V'_n$ lie, respectively, in some intervals with lengths

$$P^{1-1/n}, \ldots, P^{n(1-1/n)}, \tag{4}$$

satisfies the inequality

$$E < 2e^{r(n)} H^{\frac{n(n-3)}{2}} P^{\frac{3n-1}{2}}.$$

Proof. We shall first estimate E_1. Let s be an integer satisfying the condition $1 < s \le n$. If, for given $v_{s+1}, \ldots, v_n$, the sums $v_1, \ldots, v_n$ lie, respectively, in intervals of lengths (3), the the sums $v_1 + \ldots + v_s, \ldots, v_1^s + \ldots + v_s^s$ lie, respectively, in intervals of lengths $1, \ldots, P^{s-1}$.

Let $\eta_1, \ldots, \eta_{s\cdot}$ and $\eta_1 + \xi_1, \ldots, \eta_s + \xi_s$ be two systems of numbers $v_1, \ldots, v_s$ with the same properties and with the smallest numbers of η_s (respectively, $\xi_s > 0$). We find that

$$\frac{(\eta_1 + \xi_1) - \eta_1}{\xi_1} \xi_1 + \ldots + \frac{(\eta_s + \xi_s) - \eta_s}{\xi_s} \xi_s = 0_0,$$

$$\cdots\cdots\cdots\cdots\cdots\cdots\cdots\cdots\cdots\cdots\cdots\cdots\cdots\cdots$$

$$\frac{(\eta_1 + \xi_1)^s - \eta_1^s}{s\xi_1} \xi_1 + \ldots + \frac{(\eta_s + \xi_s)^s - \eta_s^s}{s\xi_s} \xi_s = \frac{0_{s-1}}{s} P^{s-1},$$

from which we see that

$$\Delta \xi_s - \Delta' = 0;$$

$$\Delta = \begin{vmatrix} \dfrac{(\eta_1 + \xi_1) - \eta_1}{\xi_1} & \cdots & \dfrac{(\eta_s + \xi_s) - \eta_s}{\xi_s} \\ \cdots\cdots\cdots\cdots\cdots & & \\ \dfrac{(\eta_1 + \xi_1)^s - \eta_1^s}{s\xi_1} & \cdots & \dfrac{(\eta_s + \xi_s)^s - \eta_s^s}{s\xi_s} \end{vmatrix}, \tag{5}$$

$$\Delta' = \begin{vmatrix} \dfrac{(\eta_1 + \xi_1) - \eta_1}{\xi_1} & \cdots & \dfrac{(\eta_{s-1} + \xi_{s-1}) - \eta_{s-1}}{\xi_{s-1}} 0_0 \\ \cdots\cdots\cdots\cdots\cdots\cdots\cdots\cdots\cdots\cdots\cdots\cdots \\ \dfrac{(\eta_1 + \xi_1)^s - \eta_1^s}{s\xi_1} & \cdots & \dfrac{(\eta_{s-1} + \xi_{s-1})^s - \eta_{s-1}^s 0_{s-1}}{s\xi_{s-1}} \dfrac{P^{s-1}}{s} \end{vmatrix}.$$

Next, to solve (5) we apply the following transformation. We decompose both determinants in it by the elements of the first column, and, considering the result as the difference of the values of some functions from v_1 for $v_1 = \eta_1 + \xi$, and for $v_1 = \eta_1$, we apply Lagrange's formula. We obtain a new inequality, where the elements of the first column are replaced, respectively, with the numbers $1, \ldots,$ x_1^{s-1} for some x_1 satisfying the condition $X_1 < x_1 < Y_1$. Carrying out the analogous transformation with respect to the second, third, and up to the last column and still with respect to the last column, not only for the first determinant, we obtain

$$\Delta_s \xi_s - \Delta'_s = 0,$$

$$\Delta_s = \begin{vmatrix} 1 \dots \dots 1 \\ \cdots \cdots \cdots \\ x_1^{s-1} \dots x_s^{s-1} \end{vmatrix}, \quad \Delta'_s = \begin{vmatrix} 1 \dots 1 \; \theta_0 \\ \cdots \cdots \cdots \cdots \cdots \\ x_1^{s-1} \dots x_{s-1}^{s-1} \; \dfrac{\theta_{s-1}}{s} \, P^{s-1} \end{vmatrix},$$

$$X_1 < x_1 < Y_1, \dots, X_s < x_s < Y_s.$$

From this we find that

$$\Delta'_s = \sum_{r=0}^{s-1} \frac{\theta_r}{r+1} P^r U_r,$$

where U_r is the coefficient of x_s^r in the decomposition of

$$\Delta_s = (x_s - x_1) \dots (x_s - x_{s-1}) \Delta_{s-1}$$

in powers of x_s, and, consequently, equals the product of Δ_{s-1} by the sum of the products of the numbers $-x_1, \dots, -x_{s-1}$, taken $s - 1 - r$ at a time. Therefore, we have

$$U_r \leq \Delta_{s-1} \binom{s-1}{r} P^{s-1-r},$$

$$\xi_s < \sum_{r=1}^{s-1} \frac{\binom{s-1}{r} P^{s-1}}{(r+1)(x_s - x_1) \dots (x_s - x_{s-1})}.$$

for which, because of the inequality $x_{j+1} - x_j \geq (2t - 1)R$ for $t \geq 1$, we obtain

$$\xi_s < \sum_{r=1}^{s} \frac{\binom{s}{r} H^{s-1}}{1 \cdot 3 \dots (2s-3)s} < \frac{(2^{s+1} - 2) H^{s-1}}{3 \dots (2s-1)} < L_s H^{s-1} - 1;$$

$$L_s = \frac{4}{(2 - 0,5) \dots (s - 0,5)},$$

where, since

$$\ln(2 - 0,5) + \dots + \ln(s - 0,5) > \int_1^s \ln x \, dx = s \ln s - s + 1,$$

we have

$$L_s < 4 e^{s-1} s^{-s}.$$

It follows from what has been proved so far that v_s for $s > 1$ and for given $v_{s+1}, \dots, v_n$ has only less than $4 e^{s-1} s^{-s} H^{s-1}$ different values. Since, for given $v_2, \dots, v_n$, the number v_1 lies in an interval of length 1, and, consequently, can take on no more than two different values, then we have

$$E_1 < 2 \prod_{s=2}^{n} (4e^{s-1} s^{-s} H^{s-1}) = 2 \cdot 4^{n-1} (eH)^{n(n-1)/2} \prod_{s=2}^{n} s^{-s},$$

Therefore,

$$\sum_{s=2}^{n} s \ln s > \int_{1}^{n} s \ln s \, ds > \frac{n^2}{2} \ln n - \frac{n^2}{4},$$

and so

$$E_1 < e^{r(n)-1} H^{n(n-1)/2}.$$

Since

$$\left(\frac{P^{1-1/n}}{1} + 1 \right) \cdots \left(\frac{P^{(n-1)(1-1/n)}}{P^{n-1}} + 1 \right) < eP^{(n-1)/2},$$

for integers E', the system of numbers $v_1, \ldots, v_n$ such that the sums $V_1, \ldots, V_n$ lie, respectively, in intervals of length (4), we obtain the inequality

$$E' < e^{r(n)} H^{n(n-1)/2} P^{(n-1)/2}.$$

Finally, noting that the number of all systems $v_1, \ldots, v_n$ is less than $2P^n H^{-n}$, we find

$$E < 2e^{r(n)} H^{n(n-3)/2} P^{(3n-1)/2}.$$

This completes the proof.   □

Theorem 1 (the mean value theorem of I.M. Vinogradov). *Let $\tau \geq 0$ be an integer, $k \geq n\tau$, and $P \geq 1$. Then*

$$J = J_k(P) = J_{k,n}(P) \leq D_\tau \cdot P^{2k - \Delta(\tau)},$$

where

$$\Delta(\tau) = \frac{n(n+1)}{2} \left(1 - \left(1 - \frac{1}{n} \right)^\tau \right),$$

$$D_\tau = (n\tau)^{6n\tau} (2n)^{4n(n+1)\tau}.$$

Proof. Clearly, it is sufficient to prove the Theorem for $k = n\tau$. The Theorem is true for $\tau = 1$ and any P, since the integral $J_n(P)$ equals the number of solutions of the system of equations

$$\begin{cases} x_1 + \ldots + x_n - x_{n+1} - \ldots - x_{2n} = 0, \\ \cdots\cdots\cdots\cdots\cdots\cdots\cdots\cdots\cdots\cdots \\ x_1^n + \ldots + x_n^n - x_{n+1}^n - \ldots - x_{2n}^n = 0, \end{cases}$$

$$1 \leq x_i \leq P, \quad i = 1, \ldots, 2n,$$

which do not exceed

$$n! P^n < D_1 P^{2n-n}.$$

Moreover, for $\tau \geq 1$ and $P \leq D_\tau^{1/\Delta(\tau)}$ the assertion of the Theorem is trivial. Therefore, we shall consider only the case when $\tau > 1$ and $P > D_\tau^{1/\Delta(\tau)}$.

Let m and P_0 be natural numbers, and let the Theorem hold for $\tau \leq m$, $P \leq P_0$, and also for $\tau \leq m + 1$, $P < P_0$. We shall prove that it is also true for $\tau \leq m + 1$ and $P = P_0$. Then, by the principle of mathematical induction, the Theorem is always true.

Let $k = n(m + 1)$, $H = (2n)^4$, and $R = [PH^{-1} + 1]$. Then $P \leq RH$ and $J_k(P) \leq J_k(RH)$.

We shall transform the integrand in $J_k(RH)$. First,

$$S = \sum_{\alpha=1}^{RH} e^{2\pi i f(x)} = \sum_{y=0}^{H-1} S(y),$$

where

$$S(y) = \sum_{z=1}^{R} e^{2\pi i f(z+Ry)}, \quad f(x) = \alpha_1 x + \ldots + \alpha_n x^n.$$

Consequently,

$$S^k = \sum_{y_1=0}^{H-1} \ldots \sum_{y_k=0}^{H-1} S(y_1) \ldots S(y_k).$$

A selection of integers $y_1, \ldots, y_k$, and together with it the product

$$S(y_1) \ldots S(y_k),$$

will be called *proper* if among the integers $y_1, \ldots, y_k$ there are n such that the difference of any two of them exceeds one in absolute value. Other selections of k integers and their corresponding products will be called *improper*. Now let

$$S^k = W_1 + W_2,$$

where W_1 is the sum of all proper products $S(y_1) \ldots s(y_k)$, and W_2 is the sum of all improper products. Then (by Lemma 2, Corollary 3)

$$J_k(RH) \leq 2J_1 + 2J_2,$$

where

$$J_\mu = \int_0^1 \ldots \int_0^1 |W_\mu|^2 \, d\alpha_1 \ldots d\alpha_n, \quad \mu = 1, 2.$$

We shall estimate J_1. Applying Lemma 2, Corollary 3, we find

$$J_1 \leq H^{2h} \max_{y_1, \ldots, y_k} \int_0^1 \ldots \int_0^1 |S(y_1) \ldots S(y_k)|^2 \, d\alpha_1 \ldots d\alpha_n.$$

We shall assume that the maximum is attained at the numbers $y_1, \ldots, y_k$, and that the numbers $y_1, \ldots, y_k$ are arranged so that $y_1 < y_2 < \ldots < y_k$. Moreover, $y_{v+1} - y_v > 1$ for $v = 1, 2, \ldots, n - 1$. For $v \geq n + 1$, we partition the sum

$S(y_v)$ into $\le t = [RP^{-1+1/n} + 1]$ small sums, such that for each sum the length of the interval of summation is $P^{1-1/n}$, or, perhaps, less, in the case of the last sum. Then the product

$$S(y_{n+1}) \ldots S(y_k)$$

can be presented as the sum of not more than t^{k-n} summands of the form

$$S'(y_{n+1}) \ldots S'(y_k),$$

where $S'(y_v)$ is one of the sums obtained from the partition of $S(y_v)$. Furthermore, using the fact that the geometric mean of nonnegative numbers does not exceed their arithmetic mean, we find that

$$|S'(y_{n+1})|^2 \ldots |S'(y_k)|^2 \le \frac{|S'(y_{n+1})|^{2(k-n)} + \ldots + |S'(y_k)|^{2(k-n)}}{k-n}.$$

Consequently,

$$J_1 \le t^{2(k-n)} H^{2k} \int_0^1 \ldots \int_0^1 |S(y_1) \ldots S(y_n)|^2 |S'(y)|^{2(k-n)} d\alpha_1 \ldots d\alpha_n,$$

where y is one of the numbers $y_{n+1}, \ldots, y_k$. But the last integral equals the number of solutions of the following system of equations:

$$(z_1 + Ry_1)^v + \ldots + (z_n + Ry_n)^v - (z_{n+1} + Ry_1)^v - \ldots - (z_{2n} + Ry_n)^v$$

$$= (z_{2n+1} + a)^v + \ldots - (z_{2k} + a)^v, \quad v = 1, 2, \ldots, n,$$

in which $y_1, \ldots, y_n$, a are fixed integers, $0 \le a = A + Ry < P$, $y_{\mu+1} - y_\mu > 1$ for $\mu = 1, 2, \ldots, n - 1$. Moreover, the unknowns $z_1, \ldots, z_{2n}$ range from 1 to R, and the unknowns $z_{2n+1}, \ldots, z_{2k}$ range from 1 to $P' \le P^{1-1/n}$. This system is equivalent to one of the form (Lemma 1, f):

$$(z_1 + Ry_1 - a)^v + \ldots + (z_n + Ry_n - a)^v$$

$$- (z_{n+1} + Ry_1 - a)^v - \ldots - (z_{2n} + Ry_n - a)^v = z_{2n+1}^v + \ldots - z_{2k}^v,$$

$$v = 1, 2, \ldots, n.$$

Let J denote the number of solutions of the last system of equations, and let $J'(\lambda_1, \ldots, \lambda_n)$ and $J''(\lambda_1, \ldots, \lambda_n)$ denote the number of solutions of the systems

$$(z_1 + Ry_1 - a)^v + \ldots + (z_n + Ry_n - a)^v - (z_{n+1} + Ry_1 - a)^v - \ldots$$

$$\ldots - (z_{2n} + Ry_n - a)^v = \lambda_v, \quad v = 1, 2, \ldots, n$$

and

$$z_{2n+1}^v + \ldots + z_{k+n}^v - z_{k+n+1}^v - \ldots - z_{2k}^v = \lambda_v, \quad v = 1, 2, \ldots, n.$$

Then

$$J = \sum_{\lambda_1, \ldots, \lambda_n} J'(\lambda_1, \ldots, \lambda_n) J''(\lambda_1 \ldots, \lambda_n).$$

Applying Lemma 1, b), we find

$$J \le J''(0, \ldots, 0) \sum_{\lambda_1, \ldots, \lambda_n} J'(\lambda_1, \ldots, \lambda_n)$$

$$= J_{k-n}(P^{1-1/n}) \sum_{\lambda_1, \ldots, \lambda_n} J'(\lambda_1, \ldots, \lambda_n).$$

But the last sum is equal to the number of solutions of the system of inequalities

$$|(z_1 + Ry_1 - a)^v + \ldots + (z_n + Ry_n - a)^v$$

$$- (z_{n+1} + Ry_1 - a)^v - \ldots - (z_{2n} + Ry_n - a)^v| < (k - n)P^{v(1 - 1/n)},$$

$$v = 1, 2, \ldots, n.$$

Applying the second assertion of Lemma 3, we obtain

$$\sum_{\lambda_1, \ldots, \lambda_n} J'(\lambda_1, \ldots, \lambda_n) < (2k)^n 2e^{r(n)} H^{\frac{n(n-3)}{2}} P^{\frac{3n-1}{2}}.$$

Combining these estimates, we obtain

$$J_1 \le 2(2k)^n e^{r(n)} \left(RP^{-1+\frac{1}{n}} + 1 \right)^{2(h-n)} H^{2h + \frac{n(n-3)}{2}} P^{\frac{3n-1}{2}} J_{k-n}\left(P^{1-\frac{1}{n}} \right).$$

By the induction assumption, we have

$$J_{k-n}(P^{1-1n}) < D_m P^{(1-1/n)(2k-2n-\Delta(m))}.$$

Using the fact that $P > D_{m+1}^{1/\Delta(m+1)}$, we find that

$$k = n(m + 1) > \Delta(m + 1) = \frac{n(n + 1)}{2}\left(1 - \left(1 - \frac{1}{n}\right)^{m+1}\right) \le \frac{(m + 1)(n + 1)}{2}$$

and

$$P > (2n)^{8n} \quad \text{for } m < n; \qquad \Delta(m + 1) \le n(n + 1)/2$$

and

$$P > (2n)^{8(m+1)} \quad \text{for } m > n.$$

Therefore,

$$(RP^{-1+1/n} + 1)^{2(k-n)} \le P^{2(k-n)/n} H^{-2(k-n)}(1 + 2P^{-1/n} H)^{2mn}$$

$$\le 2P^{2(k-n)/n} H^{-2(k-n)};$$

$$J_1 \le 2(2k)^n e^{r(n)} \cdot 2P^{\frac{2(k-n)}{n}} H^{-2(k-n)} H^{2k + \frac{n(n-3)}{2}} P^{\frac{3n-1}{2}}$$

$$\times D_m P^{(1-1/n)(2k-2n-\Delta(m))} < \frac{1}{4}D_{m+1} P^{2k-\Delta(m+1)}.$$

Next we estimate J_2. From the integers $0, 1, \ldots, H - 1$ one can choose no more than $H^{n-1}/(n - 1)!$ increasing sequences of $n - 1$ numbers. To each of these sequences is associated $(2n - 2)^k$ sets $y_1, \ldots, y_k$. Therefore, the general number of improper sets $y_1, \ldots, y_k$ will be no more than

$$\frac{H^{n-1}}{(n-1)!}(2n - 2)^k = B.$$

Consequently, J_2 does not exceed

$$B^2 \int_0^1 \ldots \int_0^1 |S(y_1) \ldots S(y_k)|^2 \, d\alpha_1 \ldots d\alpha_n,$$

where $y_1, \ldots, y_k$ will be a set of integers for which the last integral takes on its maximum value. Again, applying the inequality between the arithmetic and geometric means, Lemma 2, and the induction hypothesis, we obtain

$$J_2 \leq B^2 \int_0^1 \ldots \int_0^1 |S(y)|^{2k} \, d\alpha_1 \ldots d\alpha_n$$

$$= B^2 J_k(R) \leq B^2 D_{m+1} R^{2k - \Delta(m+1)} < \frac{1}{4} D_{m+1} P^{2k - \Delta(m+1)}.$$

The result now follows from the estimates for the integrals J_1 and J_2.   □

§2. Estimate of a Zeta Sum

Trigonometric sums of the form

$$\sum_{n=1}^N n^{it}, \quad \sum_{a < n \leq 2a} n^{it}$$

will be called *zeta sums*.

To estimate such sums we need two lemmas.

Lemma 4. *For* $P \geq 1$,

$$\left| \sum_{x=1}^P e^{2\pi i\alpha x} \right| \leq \min\left(P, \frac{1}{2\|\alpha\|} \right).$$

Proof. We can assume that $0 < \alpha < 1$. Then

$$\left| \sum_{a=1}^P e^{2\pi i\alpha x} \right| = \frac{|e^{2\pi i\alpha P} - 1|}{|e^{2\pi i\alpha} - 1|} \leq \frac{1}{|\sin \pi\alpha|} \leq \frac{1}{2\|\alpha\|}.$$ □

Lemma 5. *Let*

$$\alpha = \frac{a}{q} + \frac{\theta}{q^2}, \quad (a, q) = 1, \quad q \geq 1, \quad |\theta| \leq 1.$$

Then for any β, $U > 0$, and $P \geq 1$, we have

$$\sum_{x=1}^{P} \min\left(U, \frac{1}{\|\alpha x + \beta\|} \right) \leq 6\left(\frac{P}{q} + 1 \right)(U + q \log q).$$

Proof. It suffices to prove that for any β_1

$$S = \sum_{\alpha=1}^{q} \min\left(U, \frac{1}{\|\alpha x + \beta_1\|} \right) \leq 6(U + q \log q).$$

We have

$$\alpha x + \beta_1 = \frac{ax + [q\beta_1]}{q} + \frac{\theta'(x)}{q^2}, \quad \theta'(x) = \theta x + \{q\beta_1\}q, \quad |\theta'(x)| < 2q.$$

The function $\|x\|$ is periodic with period 1, and so, after making the change of variables $y = ax + [q\beta_1]$, we find

$$S = \sum_{|y| \leq q/2} \min\left(U, \frac{1}{\left\| \dfrac{y}{q} + \dfrac{0''(y)}{q} \right\|} \right),$$

where $|\theta''(y)| < 2$. If $2 < |y| \leq q/2$, then

$$\left\| \frac{y}{q} + \frac{O''(y)}{q} \right\| \geq \frac{|y| - 2}{q}$$

and

$$S \leq 5U + \sum_{2 < |y| \leq q/2} \frac{q}{|y| - 2} < 6(U + q \log q).$$

For $q \leq 6$, the Lemma is trivial. This completes the proof.   □

Theorem 2. *There exist two absolute constants $c > 0$ and $\gamma > 0$ such that, for $29N \leq t$*

$$\left| \sum_{n=1}^{N} n^{it} \right| \leq cN \exp\left(-\gamma \frac{\log^3 N}{\log^2 t} \right). \tag{6}$$

Proof. Let $100 \leq M \leq N$. We shall estimate

$$S = \sum_{n=M}^{2M} n^{it}.$$

Let $a = [M^{5/11}]$, $1 \leq x, y \leq a$. We have

$$S = \sum_{n=M}^{2M} e^{it \log(n + xy)} + 2\theta a^2, \quad |\theta| \leq 1.$$

It follows that

$$|S| \le a^{-2} \sum_{n=M}^{2M} |W(n)| + 2a^2,$$

where

$$W(n) = W = \sum_{x=1}^{a} \sum_{y=1}^{a} e^{it \log\left(1 + \frac{xy}{n}\right)}.$$

Let us estimate $|W|$. Since

$$|e^{i\varphi} - 1| = 2\left|\sin\frac{\varphi}{2}\right| \le |\varphi|,$$

then for $r \ge 1$

$$e^{it \log(1 + xy/n)} = e^{itF_1(xy)} + t\theta_1 \left(\frac{a^2}{n}\right)^{r+1},$$

where

$$F_r(xy) = \sum_{m=1}^{r} \frac{(-1)^{m-1}}{m} \left(\frac{xy}{n}\right)^m, \quad |\theta_1| \le 1.$$

We define the integer r by the condition

$$r - 1 < \frac{11 \log t}{\log M} \le r.$$

Then

$$W = W_1 + 4\theta_2 a^2 M^{-\frac{1}{13}},$$

where

$$W_1 = \sum_{x=1}^{a} \sum_{y=1}^{a} e^{2\pi i(x_1 a y + \ldots + \alpha_r x^r y^r)},$$

$$\alpha_m = \frac{(-1)^{m-1}}{2\pi m} \cdot \frac{t}{n^m}, \quad m = 1, \ldots, r; \quad |\theta_2| \le 1.$$

By Lemmas 2 and 1, for any integer $k \ge 1$,

$$|W_1|^{2k} \le a^{2k-1} \sum_{x=1}^{a} \left| \sum_{y=1}^{a} e^{2\pi i(\alpha_1 xy + \ldots + \alpha_r x^r y^r)} \right|^{2k}$$

$$\le a^{2k-1} \sum_{\lambda_1, \ldots, \lambda_r} J_{k,r}(\lambda_1, \ldots, \lambda_r) \left| \sum_{x=1}^{n} e^{2\pi i(\alpha_1 \lambda_1 x + \ldots + \alpha_r \lambda_r x^r)} \right|.$$

Furthermore, by Lemmas 1, 2, and 4,

$$|W_1|^{4k^2} \leq \alpha^{4k^2-2k} \left(\sum_{\lambda_1,\ldots,\lambda_r} J_{k,r}(\lambda_1,\ldots,\lambda_r) \right)^{2k-1}$$

$$\times \sum_{\lambda_1,\ldots,\lambda_r} J_{k,r}(\lambda_1,\ldots,\lambda_r) \left| \sum_{x=1}^{a} e^{2\pi i(\alpha_1\lambda_1 x + \ldots + \alpha_r\lambda_r x^r)} \right|^{2k}$$

$$\leq a^{8k^2-4k} J_{k,r}(0,\ldots,0) \times \left| \sum_{\substack{\lambda_1,\ldots,\lambda_r \\ \mu_1,\ldots,\mu_r}} J_{k,r}(\mu_1,\ldots,\mu_r) e^{2\pi i(\alpha_1\lambda_1\mu_1 + \ldots + \alpha_r\lambda_r\mu_r)} \right|$$

$$\leq a^{8k^2-4k} J_{k,r}(0,\ldots 0) \sum_{\mu_1,\ldots,\mu_r} J_{k,r}(\mu_1,\ldots,\mu_r)$$

$$\times \min\left(2A_1, \frac{1}{\|\alpha_1\mu_1\|}\right) \ldots \min\left(2A_r, \frac{1}{\|\alpha_r\mu_r\|}\right)$$

$$\leq a^{8k^2-4k} J_{k,r}^2(0,\ldots,0) \prod_{m=1}^{r} \sum_{|\mu_m| < A_m} \min\left(2A_m, \frac{1}{\|\alpha_m\mu_m\|}\right),$$

where

$$A_m = 2ka^m, \quad m = 1,\ldots,r.$$

For integers m in the interval

$$4\frac{\log t}{\log M} \leq m \leq 8\frac{\log t}{\log M} \tag{7}$$

we estimate the sum μ_m by using Lemma 5. For the remaining numbers m we estimate the sum μ_m trivially, namely, by the value $(2A_m)^2$. For m in the interval (7), we have

$$\sigma_m = \sum_{\mu_m} \min\left(2A_m, \frac{1}{\|\alpha_m\mu_m\|}\right) \leq 6\left(\frac{2A_m}{qm} + 1\right)(2A_m + qm\ln q_m)$$

$$\leq 6(2A_m)^2 \left(\frac{1}{q_m} + \frac{1}{A_m} + \frac{qm}{4A_m^2}\right)\ln q_m,$$

where

$$\alpha_m = \frac{(-1)^{m-1}t}{2\pi mn^m} = \frac{a_m}{q_m} + \frac{\theta_m}{q_m^2},$$

$$a_m = (-1)^{m-1}, \quad q_m = \left[\frac{2\pi mn^m}{t}\right], \quad |\theta_m| \leq 1.$$

From the conditions on m we find that

$$\sigma_m \leq 400 \cdot (32)^r (2A_m)^2 t^{-2/11};$$

$$|W_1|^{4k^2} \leq a^{8k^2-4k} J_{k,r}^2(0,\ldots,0) \cdot (400)^r (32)^{r^2} (4k)^{2r} a^{r^2+r} t^{-\frac{8}{11}\frac{\log t}{\log M}}$$

Since $a \leq M^{5/11}$, then we choose the smallest integer τ satisfying the condition

$e^t \geq 380^r$ and we apply Theorem 1 for $k = \gamma\tau$ to the estimate of $J_{k,r}(0, \ldots, 0)$ to obtain

$$|W_1|^{4k^2} \leq a^{8k^2}(400)^r(32)^{r^2}(4k)^{2r}(2r\tau)^{10r^2\tau}t^{-\frac{4}{11}\frac{\log t}{\log M}};$$

$$|W_1| \leq c_1 a^2 \exp\left(-\gamma_1 \frac{\log^3 M}{\log^2 t}\right),$$

where $c_1 > 0$ and $\gamma_1 > 0$ are absolute constants.

This gives the required estimate for $|S|$, and from this, the estimate of Theorem (6). This completes the proof.   □

Corollary. *For* $|t| \geq 2$

$$\zeta(1 + it) = O(\log^{2/3}|t|).$$

The proof follows from Theorem 6 of Chapter IV and from Lemma C of Chapter I.   □

§3. Estimate for the Zeta Function Close to the Line σ=1

It is possible to obtain an estimate in a neighborhood of the line Re $s = 1$ that is similar to the estimate found in the Corollary to Theorem 2.

Theorem 3. There exists an absolute constant $\gamma_1 > 0$ such that, for $\sigma \geq 1 - \gamma_1 \log^{-2/3}|t|$, $|t| \geq 2$, we have the estimate

$$\zeta(\sigma + it) = O(\log^{2/3}|t|).$$

Proof. We can assume that $\sigma \leq 2$. Let $\gamma_1 = \gamma/2$, where $\gamma > 0$ is the absolute constant of Theorem 2. Let $N = [\exp(\ln^{2/3}|t|]$ and $x = |t|$. By Lemma 4 of Chapter IV,

$$\zeta(s) = \sum_{n \leq N} \frac{1}{n^s} + \sum_{N < n \leq x} \frac{1}{n^s} + O(1).$$

The absolute value of the first sum does not exceed

$$\sum_{n \leq N} \frac{1}{n^\sigma} \leq 1 + \int_1^N \frac{du}{u^\sigma} = 1 + \int_1^N \frac{u^{1-\sigma}}{u} du = O(\ln N) = O(\ln^{2/3}|t|).$$

To estimate the second sum we apply Lemma C of Chapter I, where we set

$$c_n = n^{-it}, \quad C(u) = \sum_{N < n \leq u} n^{-it}, \quad f(u) = u^{-\sigma},$$

From Theorem 2, we obtain

$$\left| \sum_{N < n \leq x} \frac{1}{n^s} \right| \leq \sigma \int_N^x |C(u)| u^{-1-\sigma} du + |C(x)| x^{-\sigma}$$

$$= O\left(\int_N^x u^{-\sigma} \exp\left(-\frac{\gamma \log^3 u}{\log^2 |t|} \right) du \right) + O(|t|^{1-\sigma-\gamma})$$

$$= O\left(\int_{\log N}^{\log x} \exp\left(v(1-\sigma) - \frac{\gamma v^3}{\log^2 |t|} \right) dv \right) + O(1)$$

$$= O\left(\int_{\log N}^{\log x} \exp\left(-\frac{\gamma}{2} \frac{v^3}{\log^2 |t|} \right) dv \right) + O(1) = O(\log^{2/3} |t|).$$

This proves the Theorem.   □

§4. A Function-Theoretic Lemma

We shall need the following auxiliary result in order to refine the boundary for the zeros of the zeta function.

Lemma 6. *Let $F(s)$ be a function analytic in the circle $|s - s_0| \leq r$, let $F(s_0) \neq 0$, and let*

$$\left| \frac{F(s)}{F(s_0)} \right| \leq M$$

inside the circle. If $F(s) \neq 0$ in the domain $|s - s_0| \leq r/2$, $\mathrm{Re}(s - s_0) \geq 0$, then

a) $\mathrm{Re}\dfrac{F'(s_0)}{F(s_0)} \geq -\dfrac{4}{r} \log M$;

b) $\mathrm{Re}\dfrac{F'(s_0)}{F(s_0)} \geq -\dfrac{4}{r} \log M + \mathrm{Re}\dfrac{1}{s_0 - \rho}$,

where ρ is any zero of $F(s)$ in the domain $|s - s_0| \leq r/2$, $\mathrm{Re}(s - s_0) < 0$.

Proof. We consider the function

$$g(s) = F(s) \prod_{\rho} (s - \rho)^{-1}, \quad s \neq \rho, \quad g(\rho) = \lim_{s \to \rho} g(s),$$

where ρ runs over the zeros of $F(s)$ in the circle $|s - s_0| \leq r/2$ with multiplicity, and $g(s)$ is a function analytic in the circle $|s - s_0| \leq r$. In a neighborhood of $|s - s_0| = r$, we have

$$\left| \frac{g(s)}{g(s_0)} \right| = \left| \frac{F(s)}{F(s_0)} \prod_{\rho} \frac{s_0 - \rho}{s - \rho} \right| \leq M.$$

Consequently, the same inequality holds in the disk $|s - s_0| \leq r$. We consider the smaller circle $|s - s_0| \leq r/2$, in which $g(s) \neq 0$. Therefore, we see that $f(s) = \ln \dfrac{g(s)}{g(s_0)}$ is an analytic function in the same circle, and

$$\operatorname{Re} f(s) = \log \left| \frac{g(s)}{g(s_0)} \right| \leq \log M.$$

($M \geq 1$ by the maximum principle, since $\dfrac{g(s)}{g(s_0)} = 1$ for $s = s_0$ and $\operatorname{Re} f(s_0) = 0$.)

Therefore, applying Lemma 4, a) of Chapter II, we obtain

$$|f'(s_0)| = \left| \frac{g'(s_0)}{g(s_0)} \right| \leq \frac{4}{r} \log M;$$

$$\left| \frac{g'(s_0)}{g(s_0)} \right| = \left| \frac{F'(s_0)}{F(s_0)} - \sum_\rho \frac{1}{s_0 - \rho} \right| \leq \frac{4}{r} \log M,$$

i.e.

$$\operatorname{Re} \left\{ \frac{F'(s_0)}{F(s_0)} - \sum_\rho \frac{1}{s_0 - \rho} \right\} \geq -\frac{4}{r} \log M.$$

Since $\operatorname{Re}(s_0 - \rho) > 0$, it follows that

$$\operatorname{Re} \frac{F'(s_0)}{F(s_0)} \geq -\frac{4}{r} \log M + \operatorname{Re} \sum_\rho \frac{1}{s_0 - \rho}.$$

This completes the proof of the Lemma.

§5. A New Boundary for the Zeros of the Zeta Function

We shall now refine the results of Chapter IV, §3.

Theorem 4. *There exists an absolute constant $c > 0$ such that $\zeta(s) \neq 0$ in the domain*

$$\sigma \geq 1 - \frac{c}{\ln^{2/3}(|t| + 10) \ln \ln(|t| + 10)}.$$

Proof. Let $t \geq t_0 > 0$, where t is the ordinate of the zero $\rho = \sigma + it$. We set

$$\sigma = 1 - \frac{d}{\ln^{2/3}(2t + 2) \ln \ln(2t + 2)}, \qquad d \leq 1.$$

We must prove that $d \geq c_1 > 0$. We shall assume that t_0 is sufficiently large that

$$\frac{1}{\ln \ln(2t + 2)} < \frac{\gamma_1}{10},$$

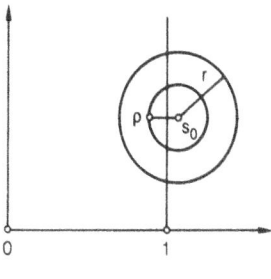

Fig. 8

where $\gamma_1 > 0$ is the constant of Theorem 3. Then

$$\frac{d}{\ln \ln (2t + 2)} < \frac{\gamma_1}{10}.$$

We consider the point

$$s_0 = 1 + \frac{4d}{\ln^{2/3}(2t + 2) \ln \ln (2t + 2)} + it = \sigma_0 + it$$

(cf. Fig. 8). We draw a circle around the point s_0 of radius r, where

$$r = \frac{\gamma_1}{\ln^{2/3}(2t + 2)}.$$

The point ρ will lie inside the circle of radius $r/2$ and center s_0 because

$$\frac{\gamma_1}{2 \ln^{2/3}(2t + 2)} > \frac{5d}{\ln^{2/3}(2t + 2) \ln \ln (2t + 2)}.$$

Setting $F(s) = \zeta(s)$ in Lemma 6, we shall estimate

$$\left| \frac{\zeta(s)}{\zeta(s_0)} \right|$$

in the circle $|s - s_0| \leq r$. By Theorem 3, in the circle $|s - s_0| \leq r$ we have

$$\zeta(s) = O(\log^{2/3} t).$$

Moreover,

$$\frac{1}{|\zeta(s_0)|} \leq \sum_{n=1}^{\infty} \frac{1}{\sigma_0} \leq 1 + \int_1^{\infty} \frac{du}{u^{\sigma_0}} = \frac{\ln^{2/3}(2t + 2) \ln \ln (2t + 2)}{4d} + 1.$$

Therefore,

$$\left| \frac{\zeta(s)}{\zeta(s_0)} \right| \leq M = c_2 \frac{\log^2 t}{d}.$$

Exactly the same estimate holds in the circle $|s - s_1| \leq r$, $s_1 = \sigma_0 + i2t$. Since

$\zeta(s) \neq 0$ in the domains $|s - s_0| \leq r/2$, $\mathrm{Re}(s - s_0) \geq 0$, and $|s - s_1| \leq r/2$, $\mathrm{Re}(s - s_1) \geq 0$, then it follows from Lemma 6 that

$$\mathrm{Re}\frac{\zeta'(s_0)}{\zeta(s_0)} \geq -\frac{4}{r}\log M + \mathrm{Re}\frac{1}{s_0 - \rho}$$

$$= -\frac{4}{\gamma_1}\ln^{2/3}(2t + 2)\ln M + \frac{\ln^{2/3}(2t + 2)\ln\ln(2t + 2)}{5d},$$

$$\mathrm{Re}\frac{\zeta'(s_1)}{\zeta(s_1)} \geq -\frac{4}{r}\log M = -\frac{4}{\gamma_1}\ln^{2/3}(2t + 2)\ln M.$$

Moreover, for $\sigma_0 > 1$,

$$-\frac{\zeta'(\sigma_0)}{\zeta(\sigma_0)} < \frac{1}{\sigma_0 - 1} + c_3.$$

Furthermore, as in Chapter IV, §2, we have

$$3\left\{-\frac{\zeta'(\sigma_0)}{\zeta(\sigma_0)}\right\} + 4\left\{-\mathrm{Re}\frac{\zeta'(\sigma_0 + it)}{\zeta(\sigma_0 + it)}\right\} + \left\{-\mathrm{Re}\frac{\zeta'(\sigma_0 + i2t)}{\zeta(\sigma_0 + i2t)}\right\} \geq 0.$$

Inserting our estimates into the last inequality, we find

$$-\frac{\ln\ln(2t + 2)}{20d} - \frac{20}{\gamma_1}\ln d + \frac{40}{\gamma_1}\ln\ln(2t + 2) + c_4 \geq 0,$$

or

$$-\frac{1}{d}\left(\frac{\ln\ln(2t + 2)}{20} + \frac{20d}{\gamma_1}\ln d\right) + \left(\frac{40}{\gamma_1}\ln\ln(2t + 2) + c_4\right) \geq 0.$$

Since $d\ln d \to 0$ and $1/d \to \infty$ as $d \to 0$, we see from the last inequality that $d \geq c_1 > 0$. The assertion of the Theorem now follows from Theorem 5, Chapter IV.   □

§6. A New Remainder Term in the Prime Number Theorem

The following is a simple consequence of Theorem 4 and the results of Chapter V, §3.

Theorem 5. *The following asymptotic formula holds for* $x \geq x_0 > 0$:

$$\psi(x) = x + O\left(x\exp\left(-c_1\left(\frac{\ln x}{\ln\ln x}\right)^{0.6}\right)\right),$$

$$\pi(x) = \int_2^x \frac{du}{\ln u} + O\left(x\exp\left(-c_2\left(\frac{\ln x}{\ln\ln x}\right)^{0.6}\right)\right).$$

Proof. Define T by $\ln T = \ln^{3/5} x \cdot (\ln \ln x)^{-3/5}$. Applying Theorem 3 of Chapter V and Theorem 4, we have

$$|\psi(x) - x| \le \left| \sum_{|\operatorname{Im} \rho| \le T} \frac{x^\rho}{\rho} \right| + O\left(\frac{x \ln^2 x}{T} \right) \le x^\sigma \sum_{|\operatorname{Im} \rho| \le T} \frac{1}{|\rho|} + O\left(\frac{x \ln^2 x}{T} \right),$$

where

$$\sigma = 1 - \frac{c}{\ln^{2/3} T \ln \ln T}.$$

This gives the first assertion of the Theorem. The second assertion follows from the first (cf. Chapter V, §2).

Exercises

1. Let $c > 0$ be an arbitrary fixed number, γ a constant satisfying $1 < \gamma < 3/2$, $m \ne 0$, $P \ge 1$,

$$f(x) = e^{c(\log x)^\gamma}, \quad S = \sum_{x=1}^{P} c^{2\pi i m f(x)}.$$

If $0 < |m| < e^{(\log P)^{3 - 2\gamma - \varepsilon}}$, where $0 < \varepsilon < 3 - 2\gamma$, then

$$|S| < c_1 P e^{-c_2(\log P)^{3 - 2\gamma}}.$$

2. Let $0 < \sigma \le 1$ and let $D(\sigma)$ be the number of numbers in the sequence $x = 1, 2, \ldots, P$ that satisfy the condition $\{f(x)\} < \sigma$. Suppose that

$$D(\sigma) = \sigma P + \lambda(\sigma).$$

If $f(x)$ satisfies the conditions of exercise 1, then

$$\lambda(\sigma) = O(P e^{-c_2(\log P)^{3 - 2\gamma}}).$$

3. Let $0.5 \le \operatorname{Re} s \le 1$, $|t| \ge 2$. Then

$$\zeta(s) = O(|t|^{c(1 - \sigma)^{3/2}} \log|t|).$$

4. For $x \ge 1$ we have

$$\sum_{n \le X} \tau_k(n) = X P_{k-1}(\log X) + O X^{1-\rho}(c_1 \log X)^k,$$

where $\rho = c/k^{2/3}$, $|\theta| \le 1$, and $P_{k-1}(u)$ is a polynomial of degree $k - 1$.

5. Let $\alpha_1, \ldots, \alpha_N$ be real numbers linearly independent over the field of rational numbers, and let $0 < \varepsilon < 1/4$. Then for any real numbers $\beta_1, \ldots, \beta_N$ we can find t such that

$$\|\alpha_1 t - \beta_1\| < \varepsilon, \ldots, \|\alpha_N t - \beta_N\| < \varepsilon.$$

6. Let

$$\Phi(X;s,\vec{\theta}) = \sum_{n\leq X} n^{-s} e^{2\pi i \sum_p \alpha_p(n)\theta_p},$$

where $n = \prod_p p^{\alpha_p(n)}$ is the canonical decomposition of n into prime factors, and the θ_p are independent real variables, indexed by the prime numbers. If

$$\Phi(X;s_0,\vec{\theta}) = 0,$$

then for any $\delta > 0$ there is an s_1 such that $\operatorname{Re} s_1 > \operatorname{Re} s_0 - \delta$ and

$$\Phi(X;s_1) = \sum_{n\leq X} \frac{1}{n^{s_1}} = 0.$$

7. For $\operatorname{Re} s > 1$ we have

$$\frac{\zeta(2s)}{\zeta(s)} = \sum_{n=1}^{\infty} \frac{\lambda(n)}{n^s} = \prod_p \left(1 - \frac{\lambda(p)}{p^s}\right)^{-1},$$

where

$$\lambda(n) = (-1)^{\sum_{p|n}\alpha_p(n)}, \quad n = \prod_p p^{\alpha_p(n)}.$$

8. Prove that

a) $\dfrac{\zeta(2s)}{\zeta(s)}\bigg|_{s=1} = 0;$

b) $\dfrac{d}{ds}\dfrac{\zeta(2s)}{\zeta(s)}\bigg|_{s=1} > 0.$

9. Let

$$F(\theta_{p_1}, \theta_{p_2}, \ldots, \theta_{p_k}) = \sum_{n=1}^{k} \ln\left(1 - \frac{e^{2\pi i\theta_{p_n}}}{P_n}\right)^{-1},$$

where $\theta_{p_1}, \theta_{p_2}, \ldots, \theta_{p_k}$ are independent real variables, indexed by the prime numbers in increasing order. Prove that for any natural number m there is a $k_0 = k_0(m)$ such that for any $k \geq k_0$ the equation

$$\operatorname{Im} F(\theta_{p_1}, \ldots, \theta_{p_k}) = \pi m$$

has a real solution.

10. Prove that there exists a completely multiplicative function $\lambda'(n)$ (that is, $\lambda'(mn) = \lambda'(m)\lambda'(n)$ for any natural numbers m and n) satisfying the conditions

a) $|\lambda'(n)| = 1$;

b) the equality $\lambda'(p) = \lambda(p) = -1$ holds for all but finitely many prime numbers,

c) the function $F(s)$, defined for $\operatorname{Re} s > 1$ by the equation

$$F(s) = \sum_{n=1}^{\infty} \frac{\lambda'(n)}{n^s},$$

is meromorphic in the entire s-plane;

d) $F'(s)|_{s=1} < 0$;

e) the following relations hold:

$\alpha)$ $\sum\limits_{n \leq X} \dfrac{\lambda'(n)}{n} = O(e^{-c\sqrt{\ln X}})$,

$\beta)$ for $\sigma \geq 1$,

$$\frac{d}{d\sigma}\left(\sum\limits_{n \leq X} \frac{\lambda'(n)}{n^{\sigma}} \right) = F'(\sigma) + O(e^{-c\sqrt{\ln X}}),$$

11. For any $\sigma \in \left[1, 1 + \dfrac{1}{\ln X} \right]$, $X \geq X_0 > 0$,

the set of values of the function F_1,

$$F_1 = F_1(\vec{\theta}_p) = \sum\limits_{0,5X < p \leq X}\left(\frac{1}{p^{\sigma}} + \frac{e^{2\pi i \theta_p}}{p^{\sigma}} \right),$$

with real variables θ_p, $0,5X < p \leq X$, is a circle $K = K(R)$ of radius $R > c/\ln X$ and center at the point $(R, 0)$.

12. There exists $c_1 > 0$ such that for $X \geq X_0 > 0$ and $\sigma = 1 + c_1/\ln X$ we have

$$-\sum\limits_{n \leq X} \frac{\lambda'(n)}{n^{\sigma}} \in K = K(R).$$

13. For $X \geq X_0 > 0$ and $\sigma = 1 + c_1/\ln X$ there exists a solution to the equation

$$\Phi(X; \sigma, \vec{\theta}) - \sum\limits_{n \leq X} n^{-\sigma} \cdot e^{2\pi i \sum\limits_{p} \alpha_p(n)\theta_p} = 0$$

with real numbers $\theta_{p_1}, \theta_{p_2}, \ldots$, where $(\Phi(X; \sigma, \vec{\theta})$ is the function defined in exercise 6.

14. For $X \geq X_0 > 0$ the equation

$$\Phi(X; s) = \sum\limits_{n \leq X} \frac{1}{n^s} = 0$$

has a solution s_1 such that

$$\operatorname{Re} s_1 > 1 + \frac{c_1}{2 \ln X}.$$

Chapter VII. The Density of the Zeros of the Zeta Function and the Problem of the Distribution of Prime Numbers in Short Intervals

It follows from the asymptotic formula for $\pi(x)$ (Theorem 5 of Chapter VI) that there exists at least one prime number in every interval $(x, x + y)$, where $x > x_0 > 0$ and

$$y = x \exp\left(-c\left(\frac{\ln x}{\ln \ln x}\right)^{0.6}\right).$$

An application of the Theorem on the density distribution of the zeros of the zeta function in the critical strip enables us to obtain a much stronger result (cf. the corollary of Theorem 2).

§1. The Simplest Density Theorem

Definition. For $0 \le \sigma \le 1$ and $T \ge 2$, the functions $N(T)$ and $N(\sigma, T)$ are defined by

$$N(T) = \sum_{|\operatorname{Im} \rho| \le T} 1, \quad N(\sigma, T) = \sum_{\substack{|\operatorname{Im} \rho| \le T \\ \operatorname{Re} \rho \le \sigma}} 1;$$

in other words, $N(T)$ is the number of nontrivial zeros of the zeta function in the rectangle $|\operatorname{Im} \rho| \le T$, and $N(\sigma, T)$ is the number of zeros of the zeta function in the rectangle $|\operatorname{Im} \rho| \le T$, $\operatorname{Re} \rho \ge \sigma$.

The problem is to obtain a precise estimate for $N(\sigma, T)$. We shall first prove a lemma.

Lemma. Let $S(t)$ be a complex-valued, continuously differentible function on the interval $[t_0, t_k]$, and let

$$t_0 < t_1 < \ldots < t_{k-1} < t_k.$$

Then setting $\delta = \min_{0 \le r < k} (t_{r+1} - t_r)$, we have

$$\sum_{r=1}^{k} |S(t_r)|^2 \le \frac{1}{\delta} \int_{t_0}^{t_k} |S(t)|^2 \, dt + 2\left(\int_{t_0}^{t_k} |S(t)|^2 \, dt\right)^{1/2} \left(\int_{t_0}^{t_k} |S'(t)|^2 \, dt\right)^{1/2}.$$

Proof. We define the function $\omega_r(t)$, the characteristic function of the interval (t_r, t_{r+1}), as follows:

$$\omega_r(t) = \begin{cases} 1 \text{ if } t_r \le t \le t_{r+1} \\ 0 \text{ otherwise.} \end{cases}$$

We set

$$\varphi_r(t) = \frac{1}{t_{r+1} - t_r} \int_{t_0}^{t} \omega_r(u)\, du.$$

Then

$$\int_{t_r}^{t_{r+1}} \varphi_r(t)(|S(t)|^2)'\, dt = \varphi_r(t)|S(t)|^2 \Big|_{t^r}^{t_{r+1}} - \frac{1}{t_{r+1} - t_r} \int_{t_r}^{t_{r+1}} |S(t)|^2 \omega_r(t)\, dt$$

$$= |S(t_{r+1})|^2 - \frac{1}{t_{r+1} - t_r} \int_{t_r}^{t_{r+1}} |S(t)|^2\, dt,$$

$$|S(t_{r+1})|^2 \le \frac{1}{\delta} \int_{t_r}^{t_{r+1}} |S(t)|^2\, dt + 2 \int_{t_r}^{t_{r+1}} |S(t)||S'(t)|\, dt.$$

We sum both sides of the inequality over r, and to the integral of product we apply Cauchy's inequality (the square of the integral of a product of nonnegative functions is not greater than the product of the integrals of the squares of the functions). This completes the proof of the Lemma.   □

Theorem 1. *For $1/2 \le \sigma \le 1$ we have the estimate*

$$N(\sigma, T) \le cT^{4\sigma(1-\sigma)}(\log T)^{12}.$$

Proof. Let $T \ge 2$. Let $x = T$ in Theorem 6 of Chapter IV. Then for $1/2 \le \sigma \le 1, |t| \le T$

$$\zeta(s) = \sum_{n \le T} \frac{1}{n^s} + \frac{T^{1-s}}{s-1} + O(T^{-\sigma} \ln T).$$

Now multiply the last equation by

$$M_X(s) = \sum_{n \le X} \frac{\mu(n)}{n^s}, \quad X = T^{2\sigma - 1},$$

This yields

$$\zeta(s)M_X(s) = \Phi(s) + R(s), \tag{1}$$

where

$$\Phi(s) = M_X(s) \sum_{n \le T} \frac{1}{n^s}, \quad R(s) = O\left(\frac{T^{1-\sigma} \ln T}{|t| + 1} |M_X(s)| \right).$$

Furthermore,

$$\Phi(s) = \sum_{m \le X} \frac{\mu(m)}{m^s} \sum_{n \le T} \frac{1}{n^s} = \sum_{n \le XT} \frac{a_n}{n^s},$$

where

$$a_n = \sum_{\substack{m \mid n \\ m \le X \le T \\ n/m \le T}} \mu(m) = \begin{cases} 1, & \text{if } n = 1; \\ 0, & \text{if } < n \le X. \end{cases} \tag{2}$$

Moreover, we always have $|a_n| \le \tau(n)$. Now let $s = \rho$, $\zeta(\rho) = 0$. Then from (1) and (2) we obtain

$$1 \le \left| \sum_{X < n \le XT} \frac{a_n}{n^\rho} \right| + O\!\left(\frac{T^{1-\sigma} \ln T}{|t| + 1} |M_X(\rho)| \right);$$

$$1 \ll \left| \sum_{X < n \le XT} \frac{a_n}{n^\rho} \right|^2 + \frac{T^{2-2\sigma} \ln^2 T}{|t|^2 + 1} |M_X(\rho)|^2.$$

Summing both sides of the last inequality over all zeros of the zeta-function in the rectangle $\sigma \le \operatorname{Re} \rho \le 1$, $|\operatorname{Im} \rho| \le T$, we find

$$N(\sigma, T) \ll \sum_\rho \left\{ \left| \sum_{X < n \le XT} \frac{a_n}{n^\rho} \right|^2 + \frac{T^{2-2\sigma} \ln^2 T}{|t|^2 + 1} |M_X(\rho)|^2 \right\}.$$

We transform the sum over ρ so that we can apply the Lemma. Let $A = [\ln T]$, and partition the inverval $[-T, \to T]$ into intervals of length of the form

$$Am + n, \quad n = 1, \dots, A; \quad |m| < TA^{-1} + 1.$$

Then

$$\sum_\rho = \sum_{|m| < TA^{-1}+1} \sum_{n=1}^{A} \sum_{Am+n-1 < \operatorname{Im} \rho \le Am+n} \le A \max_{1 \le n \le A} \sum_{|m| < TA^{-1}+1}$$

$$\times \sum_{Am+n-1 < \operatorname{Im} \rho \le Am+n}.$$

In each rectangle $Am + n - 1 < \operatorname{Im} \rho \le Am + n$ there are not more than $c_2 \ln T$ zeros, and so, choosing one zero from each such rectangle, we obtain not more than $c_3 \ln T$ sums. If we denote by $\sum_\rho'$ the largest of these, we find that

$$\sum_\rho \ll \ln^2 T \sum_\rho'.$$

If we divide $\sum_\rho'$ into not more than $c_4 \ln T$ sums, combining in one sum the summands for which $T_1 \le |\operatorname{Im} \rho| \le 2T_1$, $2T_1 \le T$ we find that

$$N(\sigma, T) \ll \ln^3 T \sum_\rho'' \left\{ \left| \sum_{X < n \le TX} \frac{a_n}{n^\rho} \right|^2 + \frac{T^{2-2\sigma} \ln^2 T}{T_1^2 + 1} |M_X(\rho)|^2 \right\}, \tag{3}$$

Moreover, the summation in Σ'' is carried out over the zeros of the zeta function such that $T_1 \le |\operatorname{Im} \rho| \le 2T_1 \le T$, $\sigma \le \operatorname{Re} \rho \le 1$, $|\operatorname{Im} \rho - \operatorname{Im} \rho'| \ge \ln T - 1$. We now estimate the sum

$$\sum_{\rho}'' \left| \sum_{Y < n \le 2Y} \frac{b_n}{n^\rho} \right|^2,$$

where the b_n are arbitrary numbers satisfying the condition $|b_n| \le \tau(n)$, and $Y \ge 1$ is any integer.

We have ($\rho = \sigma_r + it_r$)

$$\sum_{Y < n \le 2Y} \frac{b_n}{n^{\sigma_r}} n^{-it_r} = \sum_{Y < n < 2Y} \left(\frac{1}{n^{\sigma_r}} - \frac{1}{(n+1)^{\sigma_r}} \right)$$

$$\times \sum_{Y < m \le n} b_m m^{-it_r} + \frac{1}{(2Y)^{\sigma_r}} \sum_{Y < m \le 2Y} b_m m^{-it_r}.$$

Since $\sigma \le \sigma_r \le 1$, then

$$\sum_{\rho}'' \left| \sum_{Y < n \le 2Y} \frac{b_n}{n^\rho} \right|^2 \ll Y^{-2\sigma-1} \sum_{Y < n < 2Y} \sum_r \left| \sum_{Y < m \le n} b_m m^{it_r} \right|^2$$

$$+ Y^{-20} \sum_r \left| \sum_{Y < m \le 2Y} b_m m^{it_r} \right|^2.$$

We apply the Lemma to the sum S_n over r, $Y < n \le 2Y$. We obtain

$$S_n = \sum_r \left| \sum_{Y < m \le n} b_m m^{it_r} \right|^2 \ll \frac{1}{\ln T} \int_{T_1}^{2T_1} \left| \sum_{Y < m \le n} b_m m^{it} \right|^2 dt$$

$$+ \left(\int_{T_1}^{2T_1} \left| \sum_{Y < m \le n} b_m m^{it} \right|^2 dt \right)^{1/2} \left(\int_{T_1}^{2T_1} \left| \sum_{Y < m \le n} b_m m^{it} \log m \right|^2 dt \right)^{1/2}.$$

It remains to estimate the integral J,

$$J = \int_{T_1}^{2T_1} \left| \sum_{Y < m \le n} c_m m^{it} \right|^2 dt,$$

where $|c_m| \le \tau(m) \log m$.

Squaring the absolute value of the sum over m and integrating, we find that

$$J \ll T_1 \sum_{Y < m \le 2Y} |c_m|^2 + \sum_{Y < m < k \le 2Y} |c_m| |c_k| \frac{1}{\log \frac{k}{m}};$$

$$\sum_{Y < m \le 2Y} |c_m|^2 \ll \log^2 Y \sum_{Y < m \le 2Y} \tau^2(m) \ll Y \log^5 Y;$$

$$\sum_{Y < m < h \le 2Y} |c_m| |c_k| \frac{1}{\log \frac{k}{m}} \le \sum_{Y < m \le 2Y} \sum_{r=1}^{Y} |c_m| |c_{m+r}| \frac{m}{r}$$

$$\ll Y \sum_{r=1}^{Y} \frac{1}{r} \sqrt{\sum_{Y < m \le 2Y} |c_m|^2 \sum_{Y < m \le 2Y} |c_{m+r}|^2} \ll Y^2 \log^6 Y.$$

Hence,

$$J \ll (T_1 Y + Y^2) \log^6 Y; \quad S_n \ll (T_1 Y + Y^2) \log^6 Y;$$

$$\sideset{}{''}\sum_{\rho} \left| \sum_{Y < n \leq 2Y} \frac{b_n}{n^\rho} \right|^2 \ll (T_1 Y^{1-2\sigma} + Y^{2-2\sigma}) \log^6 Y. \tag{4}$$

Now, in inequality (3) we partition the first sum in the braces into $\ll \ln T$ sums and apply the estimate in (4). Noting that, in the case, $X < Y \leq XT$, we find that

$$\sideset{}{''}\sum_{\rho} \left| \sum_{X < n \leq XT} \frac{a_n}{n^\rho} \right|^2 \ll (T_1 X^{1-2\sigma} + (XT)^{2-2\sigma}) \ln^7 T;$$

Similarly, we partition the second sum in the braces in (3) into $\ll \ln T$ sums of the form (4). Noting that, in this case, $1 \leq Y \leq X$), we find that

$$\frac{T^{2-2\sigma}}{T_1^2 + 1} \sideset{}{''}\sum_{\rho} |M_X(\rho)|^2 \ll \frac{T^{2-2\sigma}}{T_1^2 + 1} (T_1 + X^{2-2\sigma}) \ln^7 T$$

$$\ll (T^{2-2\sigma} + (TX)^{2-2\sigma}) \ln^7 T \ll T^{4\sigma(1-\sigma)} \ln^7 T.$$

The Theorem now follows from these estimates, the choice of X, and (3).   □

§2. Prime Numbers in Short Intervals

Theorem 2. *Let* $h \geq x^{0.75} \exp(\ln^{0.8} x)$, $x \geq x_0 > 0$. *Then we have the following asymptotic formula*

$$\psi(x + h) - \psi(x) = h + O(h \exp(-\ln^{0.1} x),$$

Proof. For $2 \leq T \leq x$ (Theorem 3 of Chapter V),

$$\psi(x) = x - \sum_{|\operatorname{Im}\rho| < T} \frac{x^\rho}{\rho} + O\left(\frac{x \ln^2 x}{T}\right).$$

Consequently, since we can assume that $h \leq x$,

$$\psi(x + h) - \psi(x) = h - \sum_{|\operatorname{Im}\rho| \leq T} \frac{(x + h)^\rho - x^\rho}{\rho} + O\left(\frac{x \ln^2 x}{T}\right). \tag{5}$$

We estimate the sum over ρ. We have

$$\left| \frac{(x + h)^\rho - x^\rho}{\rho} \right| = \left| \int_x^{x+h} u^{\rho-1} \, du \right| \leq \int_x^{x+h} u^{\sigma-1} \, du \leq h x^{\sigma-1},$$

where $\sigma = \operatorname{Re}\rho$. Furthermore,

$$S = \sum_{|\operatorname{Im}\rho| T} x^\sigma = \sum_{|\operatorname{Im}\rho| \le T} \left(\log x \int_0^\sigma x^u\, du + 1 \right)$$

$$= N(T) + \log x \sum_{|\operatorname{Im}\rho| \le T} \int_0^1 x^u F(u, \sigma)\, du,$$

where

$$F(u, \sigma) = \begin{cases} 1, & \text{if } 0 \le u \le \sigma; \\ 0, & \text{if } \sigma < u \le 1. \end{cases}$$

From the definition of $F(u, \sigma)$ it follows that

$$\sum_{|\operatorname{Im}\rho| \le T} F(u, \sigma) = N(u, T).$$

We now note that, if $u \ge 1/2$, then by Theorem 1

$$N(u, T) \ll T^{4(1-u)} (\ln T)^{10}$$

If $0 \le u < 1/2$, then we use the trivial estimate (Corollary 1 of Theorem 4, Chapter IV):

$$N(u, T) \ll N(T) \ll T \ln T.$$

Moreover, by Theorem 4 of Chapter VI

$$N(u, T) = 0 \text{ only if } u > 1 - \frac{c}{\ln^{2,3}(T + 10)\ln\ln(T + 10)} = 1 - \gamma(T).$$

Taking all this into account, we are led to the estimate (assuming $x \ge 2T^4$)

$$S \ll T\ln T + \log x \int_0^{1/2} x^u T \ln T\, du + \log x \int_{1/2}^{1-\gamma(T)} x^u T^{4(1-u)}(\ln T)^{10}\, du$$

$$\ll x^{1/2} T \ln T + (xT^{-4})^{1-\gamma(T)} T^4 (\ln T)^{10} \ln x.$$

From this and from (5) we obtain

$$\frac{\psi(x + h) - \psi(x)}{h} = 1 + O\left(\frac{T\ln T}{\sqrt{x}} \right) + O\left(\left(\frac{T^4}{x} \right)^{\gamma(T)} (\ln T)^{10} \ln x \right)$$

$$+ O\left(\frac{x \ln^2 x}{Th} \right).$$

If we set, in the last relation,

$$T^4 = x \exp(-\ln^{0,8} x),$$

then we see that for

$$h \ge x^{0,75} \exp(\ln^{0,8} x)$$

the remainder term is

$$O(\exp(-\ln^{0,1} x)).$$

This completes the proof of the Theorem.   □

Corollary. *With the notation and conditions of Theorem 2, the interval* $(x, x + h)$ *contains a prime number.*

Proof. For $x \geq x_0$, we have

$$\sum_{x < p \leq x+h} \ln p = \psi(x + h) - \psi(x) + O(\sqrt{x} \ln^2 x)$$

$$= h + O(h \exp(-\ln^{0,1} x)) \geq 1.$$

Exercises

1. Let α be an arbitrary, fixed number in the interval $0 < \alpha \leq 1/4$, $t \geq t_0 > 0$, $X \leq \sqrt{t}$. Consider the following two statements:

$$\text{A. } \zeta\left(\frac{1}{2} + it\right) = O\left(t^{\alpha+\varepsilon}\right),$$

$$\text{B. } \sum_{n \leq X} n^{it} = O(\sqrt{X_t}^{\alpha+\varepsilon}).$$

Prove that the truth of one of these statements implies the truth of the other.

2. The Lindelöf hypothesis

$$\zeta\left(\frac{1}{2} + it\right) = O(|t|^\varepsilon),$$

is true if and only if each of the following conditions holds:

a) $\dfrac{1}{T} \int_1^T \left|\zeta\left(\dfrac{1}{2} + it\right)\right|^{2k} dt = O(T^\varepsilon), \quad k = 1, 2, \ldots$;

b) $\dfrac{1}{T} \int_1^T |\zeta(\sigma + it)|^{2k} dt = O(T^\varepsilon), \quad \sigma > 1/2, \quad k = 1, 2, \ldots$;

c) $\dfrac{1}{T} \int_1^T |\zeta(\sigma + it)|^{2k} dt \sim \sum_{n=1}^\infty \dfrac{\tau_h^2(n)}{n^{2\sigma}}, \quad \sigma > \dfrac{1}{2}, \quad k = 1, 2, \ldots$;

d) $\zeta^k(s) = \sum_{n \leq |t|^\delta} \dfrac{\tau_k(n)}{n^s} + O(|t|^{-\lambda}), \quad k = 1, 2, \ldots ; \sigma$

$\sigma \geq \sigma_0 > 1/2, \quad 0 < \delta < 1$ arbitrary, $\lambda = \lambda(k, \delta, \sigma_0) > 0$.

e) $T_k(X) = \sum_{n \leq X} \tau_k(n) = X P_{k-1}(\ln X) + O(X^{1/2+\varepsilon})$, $\quad k = 2, 3, \dots$.

3. Let $\varepsilon > 0$ be fixed, and assume the Lindelöf hypothesis. Prove that

$$N(\sigma, T) = O(T^{(2+\varepsilon)(1-\sigma)} \ln^c T).$$

4. a) Prove that for $N \geq N_0$ there exist prime numbers p and p' such that

$$N = p + p' + O(N^\gamma), \quad \gamma > 1/2. \tag{6}$$

b) Assuming that the conditions in exercise 3 are true, prove (6) with arbitrary $\gamma > 0$.

5. Assuming that the conditions in exercise 3 are true, prove that

$$\psi(x + h) - \psi(x) = h + O(h \exp(-\ln^{0,1} x)),$$

where $h \geq x^{0.5+\varepsilon}$.

Chapter VIII. Dirichlet L-Functions

Just as we studied the distribution of prime numbers in the sequence of natural numbers, we can pose and solve the problem of the distribution of prime numbers in an arithmetic progression with difference $k \geq 1$ and initial term l, where $1 \leq l \leq k$ and $(l, k) = 1$. This problem is important not only because it generalizes a classical result, but also because it has exceptional importance for the solution of many additive problems in prime number theory (for example, the Goldbach conjecture, discussed in Chapter X).

Fortunately, there exist multiplicative functions that enable us to extract from a given sequence of integers the subsequence of numbers belonging to an arithmetic progression of the form $kn + l, n = \ldots, -2, -1, 0, 1, 2, \ldots$, and this will enable us to use the methods already developed in Chapter V. These multiplicative functions, introduced by Dirichlet, are the characters $\chi(n)$. In this chapter, a character will always mean a Dirichlet character.

§1. Characters and their Properties

We shall first define characters modulo k, where k is a power of a prime number, and prove their fundamental properties. Characters to an arbitrary modulus will be defined in terms of characters whose moduli are prime powers, and so will preserve the fundamental properties of characters with prime power moduli.

Let $k = p^\alpha$, where $p > 2$ is a prime number and $\alpha \geq 1$. It is well known that there exists a primitive root modulo k. Let g be the smallest such primitive root. Let $(n, k) = 1$. We denote by $\mathrm{ind}\, n$ the index of n modulo k with respect to the primitive root g, i.e. the number $\gamma = \gamma(n) = \mathrm{ind}\, n$ satisfies the congruence

$$g^\gamma = n \,(\mathrm{mod}\, k).$$

Thus, the index of an integer is defined up to a summand that is a multiple of $\varphi(k)$.

Definition 1. Let $k = p^\alpha$, where $p > 2$ is a prime number and $\alpha \geq 1$. A *character modulo k* is a function $\chi(n)$ defined over the integers such that

$$\chi(n) = \chi(n; k) = \chi(n; k, m) = \begin{cases} 0, & \text{if } (n, k) > 1; \\ e^{2\pi i \frac{m\,\mathrm{ind}\, n}{\varphi(k)}}, & \text{if } (n, k) = 1 \end{cases}.$$

where m is an integer.

It is clear from the definition of a character that the function $\chi(n) = \chi(n; k, m)$ depends on the parameter m, and is periodic in m with period $\phi(m)$. Thus, there exist, in general, $\phi(m)$ characters modulo k, which can be obtained by taking $m = 0, 1, 2, \ldots, \phi(m) - 1$.

Now let $k = 2^\alpha$, where $\alpha \geq 3$. It is well known that for any odd number n there exist a pair of indices $\gamma_0 = \gamma_0(n)$ and $\gamma_1 = \gamma_1(n)$ modulo k such that

$$n \equiv (-1)^{\gamma_0} 5^{\gamma_1} \pmod{k}.$$

Thus, γ_0 and γ_1 are defined up to summands that are multiples of 2 and $2^{\alpha-2}$, respectively.

Definition 2. Let $k = 2^\alpha$, where $\alpha \geq 1$. A *character modulo k* is a function $\chi(n)$ defined over the integers by one of the following formulas:

$$\chi(n) = \chi(n; 2) = \chi(n; 2, 0, 0) = \begin{cases} 0, & \text{if } (n, 2) > 1; \\ 1, & \text{if } (n, 2) = 1, \end{cases}$$

$$\chi(n) = \chi(n; 4) = \chi(n; 4; m_0, 0) = \begin{cases} 0, & \text{if } (n, 4) > 1; \\ (-1)^{m_0 \nu_0}, & \text{if } (n, 4) = 1, \end{cases}$$

where $n \equiv (-1)^{\nu_0} \pmod{4}$, m_0 – integer

$$\chi(n) = \chi(n; 2^\alpha) = \chi(n; 2^\alpha, m_0, m_1)$$

$$= \begin{cases} 0, & \text{if } (n, 2^\alpha) > 1; \\ (-1)^{\nu_0 m_0} e^{2\pi i \frac{m_1 \nu_1}{2^{\alpha-2}}}, & \text{if } (n, 2^\alpha) = 1, \alpha \geq 3, \end{cases}$$

where m_0 and m_1 are integers.

It is clear from Definition 2 that the functions $\chi(n) = \chi(n; 2^\alpha, m_0, m_1)$ depend on the parameters m_0 and m_1, and are periodic in m_0 and m_1 with periods 2 and $2^{\alpha-2}$, respectively. Thus, there exist, in general, $\varphi(k) = \varphi(2^\alpha)$ characters modulo $k = 2^\alpha$, which can be obtained by taking $m_0 = 0, 1$ and $m_1 = 0, 1, 2, \ldots, 2^{\alpha-2} - 1$.

Since the index of a number or a system of indices of a number is periodic with period equal to the modulus, and since the index is also additive, i.e. the index of a product (resp. a system of indices or products) is equal to the sum of the indices of the factors (resp. the sum of the system of indices of the factors), we deduce the following properties of the characters $\chi(n)$:

1. The character $\chi(n)$ modulo k is periodic with period k, i.e. $\chi(n) = \chi(n + k)$.
2. $\chi(n)$ is a multiplicative function, i.e. $\chi(nm) = \chi(n)\chi(m)$.

Clearly, $\chi(1) = 1$.

Lemma 1. *There exist exactly $\varphi(k)$ characters modulo $k = p^\alpha, \alpha \geq 1$.*

Proof. We must show that no two of the $\phi(k)$ characters defined above are identical. First note that

$$\frac{1}{m} \sum_{x=0}^{m-1} e^{2\pi i \frac{ax}{m}} = \begin{cases} 0, & \text{if } a \not\equiv 0 \,(\mathrm{mod}\, m); \\ 1, & \text{if } a \equiv 0 \,(\mathrm{mod}\, m). \end{cases} \tag{1}$$

The first equation follows from the fact that the sum is equal to

$$\frac{e^{2\pi i \frac{am}{m}} - 1}{e^{2\pi i \frac{a}{m}} - 1} = 0, \quad \text{since } e^{2\pi i \frac{a}{m}} \neq 1.$$

The second equation is obvious. Furthermore, if n runs through a reduced system of residues modulo k, then $\gamma(n)$ (respectively, $\gamma_0(n)$ and $\gamma_1(n)$) runs through a complete system of residues modulo $\varphi(k)$ (resp. modulo 2 and modulo $2^{\alpha-2}$). Note that the cases $k = 2$ and $k = 4$ are trivial. Let $k = p^\alpha, p > 2$, and $m_1 \not\equiv m_2 (\mathrm{mod}\, \phi(k))$. If $\chi(n; k, m_1) = \chi(n; k, m_2)$, then we obtain the contradiction:

$$\varphi(k) = \sum_{\substack{n=1 \\ (n,k)=1}}^{k} \frac{\chi(n; k, m_1)}{\chi(n; k, m_2)} = \sum_{x=0}^{\varphi(k)-1} e^{2\pi i \frac{(m_1-m_2)x}{\varphi(k)}} = 0.$$

The case $k = 2^\alpha$ is proved similarly.   □

Definition 3. The *principal character*, denoted $\chi_0(n)$, is the character that is equal to 1 whenever $(n, k) = 1$.

It follows from properties 1–3 that $\chi_0(n) = \chi(n)$ for modulus $k = 2$, that $\chi_0(n) = \chi(mn; 4, 0)$ for modulus $k = 4$, that $\chi_0(n) = \chi (n; k, 0, 0)$ for modulus $k = 2^\alpha$, and that $\chi_0(n) = \chi(n; k, 0)$ for $k = p^\alpha, p > 2$.

The fundamental property of characters is their *orthogonality*. This is described in the following Lemma.

Lemma 2.

$$\frac{1}{\varphi(k)} \sum_{\chi \bmod k} \chi(n) = \begin{cases} 1, & \text{if } n \equiv 1 \,(\mathrm{mod}\, k); \\ 0, & \text{if } n \not\equiv 1 \,(\mathrm{mod}\, k), \end{cases}$$

where the summation runs over all $\varphi(k)$ characters modulo k, and

$$\frac{1}{\varphi(k)} \sum_{n=1}^{k} \chi(n) = \begin{cases} 1, & \text{if } \chi = \chi_0; \\ 0, & \text{if } \chi \neq \chi_0. \end{cases}$$

Proof. The proof follows from (1) and definitions 1–3.   □

The smallest period of a character $\chi(n)$ can be smaller than its modulus. An important role is played by those characters, called primitive characters, whose smallest period is equal to its modulus.

Definition 4. A non-principal character $\chi(n) = \chi(n; k, m)$ modulo $k = p^\alpha$, where $p > 2$ is prime, is called *primitive* if $(m, k) = 1$. A non-principal character $\chi(n) = \chi(n; k) = \chi(n; k, m_0, m_1)$ modulo $k = 2^\alpha$, where $\alpha \geq 3$, is called primitive if $(m_1, 2) = 1$. The non-principal character modulo 4 is primitive. All other non-principal characters modulo k are called *imprimitive*.

It follows immediately from Definition 4 that to each imprimitive character modulo $k = p^\alpha$ there corresponds a primitive character modulo $k_1 = p^\beta$, where $\beta < \alpha$.

There is a formula that establishes a connection between the values of a primitive character and the values of a *Gauss sum S*:

$$S = S(k; a, \chi) = \sum_{n=1}^{k} \chi(n) e^{2\pi i \frac{an}{k}}.$$

Lemma 3. *Let $\chi(n)$ be a primitive character modulo k. Then*

$$\tau(\bar{\chi})\chi(n) = \sum_{a=1}^{k} \bar{\chi}(a) e^{2\pi i \frac{an}{k}}, \tag{2}$$

where

$$\tau(\chi) = \sum_{a=1}^{k} \chi(a) e^{2\pi i \frac{a}{k}}, \quad |\tau(\chi)| = \sqrt{k}. \tag{3}$$

Proof. In the case $k = 4$, we can verify equations (2) and (3) directly. Let $k \neq 4$, $(n, k) = 1$. Defining m by the congruence $mn \equiv 1 \pmod{k}$, we have

$$\chi(n)\tau(\bar{\chi}) = \sum_{a=1}^{k} \bar{\chi}(a)\bar{\chi}(m) e^{2\pi i \frac{a}{k}} = \sum_{a=1}^{k} \bar{\chi}(am) e^{2\pi i \frac{a}{k}} = \sum_{a=1}^{k} \bar{\chi}(a) e^{2\pi i \frac{an}{k}}.$$

To obtain this result, we use the multiplicativity of $\bar{\chi}(n)$, the periodicity of $\bar{\chi}(n)$ and $e^{2\pi in/k}$, and the fact that, as the number a runs through a complete set of residues modulo k, the numbers am also run through a complete set of residues modulo k.

It remains to consider the case $(n, k) > 1$. The left side of (2) equals zero. If $k = p > 2$, then $(n, k) = p$, and the right side of (2) also equals zero, since $\chi \neq \chi_0$ and

$$\sum_{n=1}^{k} \chi(n) = 0.$$

Now let $k = p^\alpha$, where $\alpha > 1$, and let $n = rp$. Then

$$\sum_{\substack{a=1 \\ (v,p)-1}}^{p^\alpha} \bar{\chi}(a) e^{2\pi i \frac{arp}{p^\alpha}} = \sum_{v=1}^{p-1} \sum_{u=0}^{p^{\alpha-1}} \bar{\chi}(up^{\alpha-1} + v) e^{2\pi i \frac{vr}{p^{\alpha-1}}}.$$

We shall prove that

$$\sum_{u=0}^{p-1} \bar{\chi}(up^{\alpha-1} + v) = 0.$$

Using the periodicity and multiplicativity of $\bar{\chi}$ and the fact that $(v, p) = 1$, it will suffice to prove the equation

$$\sum_{u=0}^{p-1} \bar{\chi}(up^{\alpha-1} + 1) = 0.$$

Let $p > 2$. Then the primitive roots modulo p^{α} are the numbers of the form $g + pt$, where g is a primitive root modulo p, and t satisfies

$$(g + pt)^{p-1} = 1 + pb, \quad (b, p) = 1.$$

If γ is the index of the number $1 + up^{\alpha-1}$ modulo p^{α}, then $\gamma = (p-1)\gamma_1$, and

$$(g + pt)^{\nu} = (1 + pb)^{\nu_1} \equiv 1 + up^{\alpha-1} \pmod{p^{\alpha}}.$$

From this we find that

$$\gamma_1 = ub_1 p^{\alpha-2}, \quad bb_1 \equiv 1 \pmod{p}.$$

Therefore,

$$\bar{\chi}(up^{\alpha-1} + 1) = e^{-2\pi i \frac{\min d(1 + up^{\alpha-1})}{\varphi(p^{\alpha})}} = e^{-2\pi i \frac{mub_1}{p}},$$

where

$$(mb_1, p) = 1; \quad \sum_{u=0}^{p-1} e^{-2\pi i \frac{mub_1}{p}} = 0.$$

Let $p = 2$, $k = 2^{\alpha}$, $\alpha \geq 3$. Then the system of indices of the number $1 + u2^{\alpha-1}$ equals $0, 2^{\alpha-3}$, and so $(m_1, 2) = 1$,

$$\sum_{u=0}^{1} \bar{\chi}(1 + u \cdot 2^{\alpha-1}) = 1 + (-1)^0 e^{-2\pi i \frac{m_1 2^{\alpha-3}}{2^{\alpha-2}}} = 0.$$

This proves (2) for any n. From (2) and (1) we find that

$$\sum_{n=1}^{k} |\tau(\bar{\chi})|^2 |\chi(n)|^2 = \varphi(k)|\tau(\bar{\chi})|^2 = \sum_{n=1}^{k} \left| \sum_{n=1}^{k} \bar{\chi}(a) e^{2\pi i \frac{an}{k}} \right|^2$$

$$= \sum_{a,b=1}^{k} \bar{\chi}(a)\chi(b) \sum_{n=1}^{k} e^{2\pi i \frac{(a-b)n}{k}} = k\varphi(k),$$

which proves (3). This completes the proof of the Lemma. □

We shall now define characters for an arbitrary modulus k. Let $k = p_1^{\alpha_1} \dots p_r^{\alpha_r}$ be the canonical decomposition of k into prime factors.

Definition 5. A *character* $\chi(n)$ modulo k is a function defined by the equation

$$\chi(n) = \chi(n; k) = \prod_{t=1}^{r} \chi(n; p_t^{\alpha_t}). \tag{4}$$

Definition 6. A character modulo k is called *principal* if, in (4),

$$\chi(n; p_t^{\alpha_t}) = \chi_0(n; p_t^{\alpha_t}), \quad t = 1, \ldots, r.$$

Definition 7. A non-principal character modulo k is called *primitive* if, in (4), $\chi(n; p_t^{\alpha_t})$ is a primitive character modulo $p_t^{\alpha_t}$ for $t = 1, \ldots, r$. Otherwise, $\chi(n)$ is called *imprimitive*.

It follows from Definition 7 that to any imprimitive character $\chi(n)$ modulo k there corresponds a primitive character $\chi(n)$ modulo k_1 that is equal to $\chi(n)$ on numbers that are relatively prime to k. Moreover, k_1 divides k. In this case we say that $\chi(n)$ is the character induced by $\chi_1(n)$, and $\chi_1(n)$ is called the primitive character that induces χ.

All of the statements proved above about characters modulo $k = p^\alpha$ are also true for arbitrary k and are simple consequences of what has already been proved.

We now formulate the fundamental properties of a character $\chi(n)$ modulo k.

1. The character $\chi(n)$ modulo k is an arithmetic function that is periodic with period k, and not identically zero. Moreover, $\chi(n) = 0$ if $(n, k) > 1$ and $\chi(n) \neq 0$ if $(n, k) = 1$.

2. $\chi(n)$ is completely multiplicative, that is, $\chi(nm) = \chi(n)\chi(m)$ for all n and m.

3. There exist exactly $\varphi(k)$ different characters modulo k.

4. The orthogonality property:

$$\frac{1}{\varphi k} \sum_{\chi \bmod k} \chi(n) = \begin{cases} 1, & \text{if } n \equiv 1 \ (\mathrm{mod}\ k); \\ 0, & \text{if } n \not\equiv 1 \ (\mathrm{mod}\ k), \end{cases}$$

where the summation runs over each of the $\varphi(k)$ characters modulo k; and

$$\frac{1}{\varphi k} \sum_{n=1}^{k} \chi(n) = \begin{cases} 1, & \text{if } \chi = \chi_0; \\ 0, & \text{if } \chi \neq \chi_0. \end{cases}$$

5. Let χ be a primitive character modulo k. Then

$$\tau(\bar{\chi})\chi(n) = \sum_{m=1}^{k} \bar{\chi}(m) e^{2\pi i \frac{mn}{k}}, \tag{5}$$

where

$$\tau(\chi) = \sum_{n=1}^{k} \chi(n) e^{2\pi i \frac{n}{k}}, \quad |\tau(\bar{\chi})| = \sqrt{k}.$$

It is easy to prove properties 1–5. For example, we shall prove property 5. Let $k = k_1 k_2$, where $(k_1, k_2) = 1$. Then

$$\chi(m; k) = \chi(m; k_1)\chi(m; k_2).$$

The expression $m_1 k_2 + m_2 k_1$ runs through a complete system of residues modulo $k_1 k_2$ when m_1 and m_2 run through complete systems of residues modulo

k_1 and k_2. Therefore,

$$S = S(n, k) = \sum_{m=1}^{k} \bar{\chi}(m) e^{2\pi i \frac{mn}{k}}$$

$$= \sum_{m_1=1}^{k_1} \sum_{m_2=1}^{k_2} \bar{\chi}(m_1 k_2 + m_2 k_1; k_1) \bar{\chi}(m_1 k_2 + m_2 k_1; k_2)$$

$$\times e^{2\pi i \frac{(m_1 k_2 + m_2 k_1)n}{k_1 k_2}}$$

$$= \left(\sum_{m_1=1}^{k_1} \bar{\chi}(m_1 k_2; k_1) e^{2\pi i \frac{m_1 n}{k_1}} \right) \left(\sum_{m_2=1}^{k_2} \bar{\chi}(m_2 k_1; k_2) e^{2\pi i \frac{m_2 n}{k_2}} \right)$$

$$= \bar{\chi}(k_2; k_1) \bar{\chi}(k_1; k_2) S(n, k_1) S(n, k_2).$$

Moreover, $\tau(\chi) = S(1, k)$. Formula (5) follows from this and from Lemma 3. The character $\chi(n)$ modulo k can also be defined by properties 1 and 2.

Lemma 4. *Let $Y(n)$ be an arithmetic function that is periodic with period k, not identically zero, and completely multiplicative, i.e. $Y(nm) = Y(n) Y(m)$. Moreover, suppose $Y(n) = 0$ if $(n, k) > 1$. Then*

$$Y(n) = \chi(n; k, m)$$

for some m.

Proof. Let $(a, k) = 1$. Then

$$T = \sum_{n=1}^{k} Y(n) \bar{\chi}(n) = \sum_{n=1}^{k} Y(an) \bar{\chi}(an) = Y(a) \bar{\chi}(a) T.$$

Therefore, either $Y(a) = \chi(a)$ for some χ, or $T = 0$ for some χ. But then, for any b with $(b, k) = 1$,

$$0 = \sum_{\chi} \chi(b) \sum_{n=1}^{k} Y(n) \bar{\chi}(n) = \sum_{n=1}^{k} Y(n) \sum_{\chi} \chi(b) \chi(\bar{n}) = Y(b) \varphi(k),$$

which contradicts the hypothesis. This proves the Lemma.   □

Corollary. *The product of a character modulo k_1 and a character modulo k_2 is a character modulo $k_1 k_2$.*

Characters are complex-valued functions. An important role is played by characters that are non-principal and real-valued. There are called *real*. For example, if $p > 2$ is prime, then the following is a real character modulo p:

$$\chi(n) = \chi\left(n; p, \frac{p-1}{2}\right) = \begin{cases} 0, & \text{if } (n, p) > 1; \\ (-1)^{\text{ind } n}, & \text{if } (n, p) = 1. \end{cases}$$

This character is called the *Legendre symbol* and denoted (n/p). A character that takes even one complex value is called *complex*, and the character whose values are the complex conjugates of the values of $\chi(n)$ is called the complex conjugate of $\chi(n)$, and is denoted $\bar\chi(n)$. For every character modulo k there is the identity.

$$\chi^{\varphi(k)}(n) = \chi_0(n).$$

The smallest natural number r for which $\chi^r(n) = \chi_0(n)$ is called the *degree* of the character. Thus, the principal character has degree 1, a real character has degree 2, and a complex character has degree 3 or greater.

Because of the multiplicativity of characters, we have

$$\chi^2(-1) = 1,$$

i.e. $\chi(-1) = \pm 1$. A character for which $\chi(-1) = +1$ is called *even* and a character for which $\chi(-1) = -1$ is called *odd*.

We note another property of characters. If $\chi \neq \chi_0$ is a character modulo k, then for any M

$$\left| \sum_{n=1}^{M} \chi(n) \right| \leq \varphi(k).$$

We can improve this inequality in the case when χ is a primitive character.

Lemma 5. *Let χ be a primitive character modulo k,*

$$S = \sum_{n < M} \chi(n).$$

Then

$$|S| < \sqrt{k}\ln k.$$

Proof. We can assume that $M \leq k - 1$. By property 5,

$$\chi(n) = \frac{1}{\tau(\bar\chi)} \sum_{m=1}^{k} \bar\chi(m)\, e^{2\pi i \frac{mn}{k}},$$

and so

$$S = \frac{1}{\tau(\bar\chi)} \sum_{m=1}^{k} \bar\chi(m) \sum_{n < M} e^{2\pi i \frac{mn}{k}} = \frac{1}{\tau(\bar\chi)} \sum_{m=1}^{k-1} \bar\chi(m) \frac{e^{2\pi i \frac{mM}{k}} - 1}{e^{2\pi i \frac{m}{k}} - 1}.$$

Passing to an inequality, we find that

$$|S| \leq \frac{1}{\sqrt{k}} \sum_{m=1}^{k-1} \frac{\left| \sin \pi \frac{mM}{k} \right|}{\left| \sin \pi \frac{m}{k} \right|} < \frac{1}{\sqrt{k}} \sum_{m=1}^{k-1} \frac{1}{\left| \sin \pi \frac{m}{k} \right|}.$$

If k is an odd integer, then

$$|S| \leq \frac{2}{\sqrt{k}} \sum_{m=1}^{(k-1)/2} \frac{1}{\sin \pi \dfrac{m}{k}} \leq \sqrt{k} \sum_{m=1}^{(k-1)/2} \frac{1}{m},$$

since $\sin \pi \alpha \geq 2\alpha$ for $0 \leq \alpha \leq 1/2$.

If k is an even integer, then

$$|S| < \frac{2}{\sqrt{k}} \sum_{m=1}^{k'2-1} \frac{1}{\sin \pi \dfrac{m}{k}} + \frac{1}{\sqrt{k}} \leq \sqrt{k} \sum_{m=1}^{k'2-1} \frac{1}{m} + \frac{1}{\sqrt{k}}.$$

Moreover,

$$\frac{1}{m} \leq \ln \frac{2m+1}{2m-1}, \quad m \geq 1,$$

$$\sum_{m=1}^{(h-1)/2} \frac{1}{m} \leq \ln k, \quad k - \text{odd}$$

$$\sum_{m=1}^{k/2-1} \frac{1}{m} \leq \ln(k-1) \leq \ln k - \frac{1}{k}, \quad k - \text{even}$$

This completes the proof. □

§2. Definition of L-Functions and their Simplest Properties

Dirichlet L-functions are functions of a complex variable, similar to the Riemann zeta function, that Dirichlet introduced in order to study the problem of the distribution of prime numbers in arithmetic progressions. By an L-function we shall always mean a Dirichlet L-function.

Let k be a natural number, and let χ be a character modulo k.

Definition 8. An *L-function* is a series of the form

$$L = L(s, \chi) = \sum_{n=1}^{\infty} \frac{\chi(n)}{n^s}, \quad \text{Re } s > 1.$$

It follows from the fact that $|\chi(n)| \leq 1$ that $L(s, \chi)$ is analytic in the half-plane Re $s > 1$. There is an analog of the Euler product for $L(s, \chi)$.

Lemma 6. *For Re $s > 1$, we have*

$$L(s, \chi) = \prod_p \left(1 - \frac{\chi(p)}{p^s}\right)^{-1}. \tag{6}$$

Proof. For $X > 1$ we consider the function

$$\Phi(s; X) = \prod_{p \leq X} \left(1 - \frac{\chi(p)}{p^s}\right)^{-1}.$$

Since Re $s > 1$, we have

$$\left(1 - \frac{\chi(p)}{p^s}\right)^{-1} = 1 + \frac{\chi(p)}{p^s} + \frac{\chi(p^2)}{p^{2s}} + \cdots$$

Consequently, using the multiplicitivity of $\chi(n)$ and the unique decomposition of a natural number into prime factors, we obtain

$$\Phi(s; N) = \prod_{p \leq X} \left\{1 + \frac{\chi(p)}{p^s} + \frac{\chi(p^2)}{p^{2s}} + \cdots\right\} = \sum_{n \leq X} \frac{\chi(n)}{n^s} + R(s, X). \qquad (7)$$

Furthermore,

$$|R(s; X)| \leq \sum_{n > X} \frac{1}{n^\sigma} < \int_X^\infty \frac{du}{u^\sigma} = \frac{1}{\sigma - 1} X^{1-\sigma},$$

where $\sigma = $ Re $s > 1$. Passing to the limit in (7) as $X \to \infty$ completes the proof.
□

From (6) we find

$$\left|\frac{1}{L(s, \chi)}\right| = \left|\prod_p \left(1 - \frac{\chi(p)}{p^s}\right)\right| \leq \sum_{n-1}^\infty \frac{1}{n^\sigma} < 1 + \int_1^\infty \frac{du}{u^\sigma} = 1 + \frac{1}{\sigma - 1},$$

$$|L(s, \chi)| > \frac{\sigma - 1}{\sigma},$$

i.e. $L(s, \chi) \neq 0$ for Re $s > 1$. If the character χ modulo k is principal, then $L(s, \chi)$ differs from the zeta function $\zeta(s)$ by only a simple factor.

Lemma 7. *Let $\chi(n) = \chi_0(n)$ modulo k. Then for Re $s > 1$,*

$$L(s, \chi_0) = \zeta(s) \prod_{p \backslash k} \left(1 - \frac{1}{p^s}\right).$$

Proof. The Lemma follows from (6) and the definition of the principal character $\chi_0(n)$.

Corollary. $L(s, \chi_0)$ *is an analytic function in the entire s-plane except at the point $s = 1$, where it has a simple pole with residue equal to*

$$\prod_{p \backslash k} \left(1 - \frac{1}{p}\right).$$

If $\chi(n)$ is an imprimitive character, and $\chi_1(n)$ is the primitive character modulo k_1, where $k_1|k$, corresponding to $\chi(n)$, then $L(s, \chi)$ differs from $L(s, \chi_1)$ by only a simple factor.

Lemma 8. *Let χ_1 be a primitive character modulo k_1, and let χ be the imprimitive character modulo k induced by χ_1, where $k_1 \neq k$. Then for Re $s > 1$*

$$L(s, \chi) = L(s, \chi_1) \prod_{\substack{p \mid k \\ p \times k_1}} \left(1 - \frac{\chi_1(p)}{p^s}\right).$$

Proof. The Lemma follows from (6) and the properties of χ and χ_1. □

The function $L(s, \chi)$ can easily be continued over the plane Re $s > 0$.

Lemma 9. *Let $\chi \neq \chi_0$. Then for Re $s > 0$ we have*

$$L(s, \chi) = s \int_1^\infty S(\chi) x^{-1} dx, \tag{8}$$

where

$$S(x) = \sum_{n \leq x} \chi(n).$$

Proof. Let $N \geq 1$ and Re $s > 1$. Applying partial summation (Lemma C of Chapter I), we have

$$\sum_{n=1}^N \frac{\chi(n)}{n^s} = 1 + s \int_1^N c(x) x^{-s-1} dx + \frac{c(N)}{N^s},$$

where

$$c(x) = S(x) - 1.$$

Passing to the limit as $N \to +\infty$, we obtain (8) for Re $s > 1$. But $|S(x) \leq \varphi(k)$. Therefore, the integral in (8) converges in the half plane Re $s > 0$ and defines there an analytic function. This completes the proof. □

Corollary. *For Re $s \geq 1/2$ and $\chi \neq \chi_0$, we have the estimate*

$$|L(s, \chi)| \leq 2|s|\varphi(k).$$

By taking the logarithm of (6) and then differentiating, we obtain the following result.

Lemma 10. *For Re $s > 1$ we have*

$$\frac{L'(s, \chi)}{L(s, \chi)} = -\sum_{n=1}^\infty \frac{\Lambda(n)\chi(n)}{n^s}. \tag{9}$$

Next we apply Theorem 1 of Chapter V to (9) with $b = 1 + 1/\ln x$, $\alpha = 1$, and $T \geq 2$, and we obtain

$$\varphi(x, \chi) = \sum_{n \leq x} \Lambda(n)\chi(n) = \frac{1}{2\pi i} \int_{b-iT}^{b+iT} \left(-\frac{L'(s, \chi)}{L(s, \chi)}\right) \frac{x^s}{s} ds + O\left(\frac{x \ln^2 x}{T}\right) \tag{10}$$

Furthermore, for $(k, l) = 1$,

$$\psi(x, k, l) - \frac{1}{q(k)} \sum_{\chi \bmod k} \psi(x, \chi) \bar{\chi}(l).$$

Thus, in order to know the behavior of $\psi(x; k, l)$, it is necessary to know the behavior of $\psi(x, \chi)$ for all χ modulo k, i.e. it is necessary to know the behavior of integral in (10). In order to study the behavior of the integral in (10), it is necessary to have for $L(s, \chi)$ the same kind of information that we had obtained about the zeta function (cf. Chapters IV–V).

The study of the function $L(s, \chi)$ is similar to the study of $\zeta(s)$, but with some additional difficulties. Before continuing $L(s, \chi)$ to the entire s-plane, we must prove that a certain related function $\xi(s, \chi)$ is an entire function of order one. For this we shall need to use Theorem 5 of Chapter II.

§3. The Functional Equation

We shall obtain a functional equation for $L(s, \chi)$ with primitive character χ. From this and from Lemma 8, we shall continue $L(s, \chi)$ over the entire s-plane for any χ. The form of the functional equation will depend on whether the character is even or odd, that is, whether $\chi(-1) = +1$ or $\chi(-1) = -1$.

Before considering the functional equation for $L(s, \chi)$ and the continuation of $L(x, \chi)$ to the entire s-plane, we shall prove a useful result that is analogous to the functional equation for $\theta(x)$ (cf. Lemma 3 of Chapter IV).

Lemma 11. *Let χ be a primitive character modulo k. For an even character χ we define the function $\theta(x, \chi)$ by*

$$\theta(x, \chi) - \sum_{n=-\infty}^{+\infty} \chi(u) e^{-\frac{n^2 \pi x}{k}}, \quad x > 0.$$

For an odd character χ we define the function $\theta_1(x, \chi)$ by the equation

$$\theta_1(x, \chi) = \sum_{n=-\infty}^{+\infty} n\chi(n) e^{-\frac{n^2 \pi x}{k}}, \quad x > 0.$$

Then the following functional equations hold for the functions $\theta(x, \chi)$ and $\theta_1(x, \chi)$:

$$\tau(\bar{\chi}) 0(x, \chi) = \sqrt{\frac{k}{x}}\, 0\left(\frac{1}{x}, \bar{\chi}\right); \tag{11}$$

$$\tau(\bar{\chi}) \theta_1(x, \chi) = i \sqrt{\frac{k}{x^3}}\, \theta_1\left(\frac{1}{x}, \bar{\chi}\right), \tag{12}$$

where $\tau(\chi)$ is Gauss's sum.

Proof. We apply the equation proved in Chapter IV, Lemma 3:

$$\sum_{n=-\infty}^{+\infty} e^{-\frac{\pi(n+a)^2}{x}} = \sqrt{x} \sum_{n=-\infty}^{+\infty} e^{-\pi n^2 x + 2\pi i n a}, \tag{13}$$

where $x > 0$ and α is real.

We have

$$\tau(\bar{\chi})\theta(s, \chi) = \sum_{m=1}^{k} \bar{\chi}(m) \sum_{n=-\infty}^{+\infty} e^{-\frac{n^2 \pi x}{k} + \frac{2\pi i m n}{k}}$$

$$= \sum_{m=1}^{k} \bar{\chi}(m) \sqrt{\frac{k}{x}} \sum_{n=-\infty}^{+\infty} e^{-\frac{k\pi\left(n+\frac{m}{k}\right)^2}{x}}$$

$$= \sqrt{\frac{k}{x}} \sum_{m-1}^{k} \bar{\chi}(m) \sum_{n=-\infty}^{+\infty} e^{-\frac{\pi(kn+m)^2}{kx}}$$

$$= \sqrt{\frac{k}{x}} \sum_{n=-\infty}^{+\infty} \bar{\chi}(m) e^{-\frac{\pi m^2}{kx}} = \sqrt{\frac{k}{x}} \theta\left(\frac{1}{x}, \bar{\chi}\right),$$

which proves equation (11).

In order to prove equation (12), we differentiate equation (13) and replace x by x/k and α by m/k. We can differentiate the series in (13) term by term because the resulting series converges uniformly. We obtain

$$\sum_{n=-\infty}^{+\infty} ne^{-\frac{\pi n^2 x}{k} + \frac{2\pi i m n}{k}} = i \sqrt{\frac{k}{x^3}} \sum_{n=-\infty}^{+\infty} (kn + m)e^{-\frac{\pi(kn+m)^2}{kx}}.$$

From this, we obtain as above

$$\tau(\bar{\chi})\theta_1(x, \chi) = \sum_{m=1}^{k} \bar{\chi}(m) \sum_{n=-\infty}^{+\infty} ne^{-\frac{n^2 \pi x}{k} + \frac{2\pi i m n}{k}}$$

$$- i \sqrt{\frac{k}{x^3}} \sum_{m=1}^{k} \bar{\chi}(m) \sum_{n=-\infty}^{+\infty} (kn + m)e^{-\frac{\pi(kn+m)^2}{kx}}$$

$$- i \sqrt{\frac{k}{x^3}} \sum_{n=-\infty}^{+\infty} n\bar{\chi}(n)e^{-\frac{\pi n^2}{kx}} = i \sqrt{\frac{k}{x^3}} \theta_1\left(\frac{1}{x}, \bar{\chi}\right).$$

This proves the Lemma   □

Theorem 1 (functional equation). *Let χ be a primitive character modulo k,*

$$\delta = \begin{cases} 0, & \text{if } \chi(-1) = +1; \\ 1, & \text{if } \chi(-1) = -1; \end{cases}$$

$$\xi(s, \chi) = \left(\frac{\pi}{k}\right)^{-(s+\delta)/2} \Gamma'\left(\frac{s+\delta}{2}\right) L(s, \chi).$$

Then

$$\xi(1 - s, \bar{\chi}) = \frac{i^\delta \sqrt{k}}{\tau(\chi)} \xi(s, \chi). \tag{14}$$

Proof. The proof is essentially a repetition of the arguments used to obtain the functional equation for the zeta function (Chapter IV, Theorem 1).
Assume that $\chi(-1) = +1$. We have

$$\pi^{-\frac{s}{2}} k^{\frac{s}{2}} \Gamma\left(\frac{s}{2}\right) n^{-s} = \int_0^\infty e^{-\frac{n^2 \pi x}{k}} x^{\frac{s}{2}-1} dx.$$

Multiplying the last equation by $\chi(n)$ and summing over n, we obtain for Re $s > 1$

$$\pi^{-\frac{s}{2}} k^{\frac{s}{2}} \Gamma\left(\frac{s}{2}\right) L(x, \chi) = \int_0^\infty x^{\frac{s}{2}-1} \left(\sum_{n=1}^\infty \chi(n) e^{-\frac{n^2 \pi x}{k}} \right) dx.$$

Since χ is an even character, we have

$$\sum_{n=1}^\infty \chi(n) e^{-\frac{n^2 \pi x}{k}} = \frac{1}{2} \theta(x, \chi);$$

$$\left(\frac{\pi}{k}\right)^{-\frac{s}{2}} \Gamma\left(\frac{s}{2}\right) L(s, \chi) = \frac{1}{2} \int_0^\infty x^{\frac{s}{2}-1} \theta(x, \chi) dx.$$

Dividing the last integral into two parts, making in one of them the change of variable of integration $(x \to 1/x)$, and applying (11), we find that

$$\left(\frac{\pi}{k}\right)^{-\frac{s}{2}} \Gamma\left(\frac{s}{2}\right) L(x, \chi) = \frac{1}{2} \int_1^\infty x^{\frac{s}{2}-1} \theta(x, \chi) dx + \frac{1}{2} \int_1^\infty x^{-\frac{s}{2}-1} \theta\left(\frac{1}{x}, \chi\right) dx$$

$$= \frac{1}{2} \int_1^\infty x^{\frac{s}{2}-1} \theta(x, \chi) dx + \frac{1}{2} \frac{\sqrt{k}}{\tau(\chi)} \int_1^\infty x^{-\frac{s}{2}-\frac{1}{2}} \theta(x, \bar{\chi}) dx. \tag{15}$$

The right side of this equation is an analytic function for any s, and, consequently, gives the analytic continuation of $L(s, \chi)$ over the entire s-plane. Since $\Gamma(s/2) \neq 0$, then $L(s, \chi)$ is an everywhere regular function. Furthermore, after replacing s by $1 - s$ and χ by $\bar{\chi}$, the side of (15) is multiplied by $\sqrt{k}/\tau(\chi)$, since $\chi(-1) = 1$, and, consequently, $\tau(\chi)\tau(\bar{\chi}) = \tau(\chi)\overline{\tau(\chi)} = k$. This gives the assertion of the Theorem for $\delta = 0$.
Next we assume that $\chi(-1) = -1$. We have

$$\left(\frac{\pi}{k}\right)^{-\frac{s+1}{2}} \Gamma\left(\frac{s+1}{2}\right) n^{-s} = \int_0^\infty n e^{-\frac{\pi n^2 x}{k}} x^{\frac{s}{2}-\frac{1}{2}} dx.$$

Consequently, for Re $s > 1$,

$$\left(\frac{\pi}{k}\right)^{-\frac{s+1}{2}}\Gamma\left(\frac{s+1}{2}\right)L(s,\chi) = \frac{1}{2}\int_0^\infty \theta_1(x,\chi)x^{\frac{s}{2}-\frac{1}{2}}dx$$

$$= \frac{1}{2}\int_1^\infty \theta_1(x,\chi)x^{\frac{s}{2}-\frac{1}{2}}dx + \frac{i\sqrt{k}}{2\tau(\bar\chi)}\int_1^\infty \theta_1(x,\bar\chi)x^{-\frac{s}{2}}dx.$$

The last equation gives the regular continuation of $L(s,\chi)$ over the entire s-plane. If we replace s by $1-s$ and χ by $\bar\chi$, the right side is multiplied by $i\sqrt{k}\tau(\chi)$, since

$$\tau(\chi)\tau(\bar\chi) = -k.$$

This proves the Theorem for $\delta = 1$. □

Corollary. $\xi(s,\chi)$ *is an entire function. If* $\chi(-1) = +1$, *then the only zeros of* $L(s,\chi)$ *for Re* $s \le 0$ *are the poles of* $\Gamma(s/2)$, *i.e. the points* $s = 0, -2, -4, \ldots$. *If* $\chi(-1) = -1$, *then the only zeros of* $L(s,\chi)$ *for Re* $s \le 0$ *are the poles of* $\Gamma((s+1)/2)$, *i.e. the points* $s = -1, -3, -5, \ldots$.

We shall prove below (cf. Chapter IX, §2) that $L(1,\chi) \ne 0$. It follows from this and (14) that $\xi(0,\chi) \ne 0$.

§4. Non-trivial Zeros; Expansion of the Logarithmic Derivative as a Series in the Zeros

It is clear from the Corollary to Theorem 1 that the function $L(s,\chi)$, where χ is a primitive character, has only real zeros in the half-plane Re $s < 0$. These zeros are the poles of either $\Gamma(s/2)$ or $\Gamma((s+1)/2)$, and are called the trivial zeros. The zero $s = 0$ is also called trivial zero. Besides the trivial zeros, the function $L(s,\chi)$ has, like the zeta-function, infinitely many non-trivial zeros lying in the strip $0 \le$ Re $s \le 1$ (the critical strip).

Theorem 2. *Let* χ *be a primitive character. Then the function* $\xi(s,\chi)$ *is an entire function of order one with infinitely many zeros* ρ_n *such that* $0 \le$ Re $\rho_n \le 1, \rho_n \ne 0$. *Moreover, the series* $\sum\limits_{n=1}^\infty |\rho_n|^{-1}$ *diverges, but the series* $\sum\limits_{n=1}^\infty |\rho_n|^{-1-\varepsilon}$ *converges for any* $\varepsilon > 0$. *The zeros of* $\xi(s,\chi)$ *are the non-trivial zeros of* $L(s,\chi)$.

Proof. For Re $s \ge 1/2$,

$$|L(s,\chi)| \le 2|s|\varphi(k) < 2|s|k;$$

$$|\xi(s,\chi)| \le 2k^{\frac{\sigma}{2}+\frac{3}{2}}|s|\left|\Gamma\left(\frac{s+\delta}{2}\right)\right| \ll k^{\frac{\sigma}{2}+\frac{3}{2}}e^{c|s|\ln|s|}.$$

Because of the functional equation (14) and the fact that

$$\left| \frac{i^\delta \sqrt{k}}{\tau(\chi)} \right| = 1$$

the last estimate for $|\xi(s, \chi)|$ also holds for Re $s < 1/2$. Moreover, $\xi(0, \chi) \neq 0$. Since $\ln \Gamma(s) \sim s$ as $s \to +\infty$, the first assertion of the Theorem follows from Chapter II, Theorem 5. Since $L(s, \chi) \neq 0$ for Re $s > 1$, then it follows from (14) that $\xi(s, \chi) \neq 0$ for Re $s < 0$, i.e. the zeros of $\xi(s, \chi)$ are the non-trivial zeros of $L(s, \chi)$ that lie in the strip $0 \leq$ Re $s \leq 1$. This proves the Theorem. □

Corollary. *The following formula holds*

$$\xi(s, \chi) = e^{A + Bs} \prod_{n=1}^{\infty} \left(1 - \frac{s}{\rho_n} \right) e^{\frac{s}{\rho_n}}, \tag{16}$$

where $A = A(\chi)$ and $B = B(\chi)$ are constants.

It follows from (14) that the non-trivial zeros of $L(s, \chi)$ are symmetric with respect to the line Re $s = 1/2$. We shall assume below that the zeros ρ_n, $n = 1, 2, \ldots$, are enumerated in order of the increasing absolute value of their imaginary parts.

The following useful statement establishes a connection between the constant $B = B(\chi)$ and the non-trivial zeros of $L(s, \chi)$.

Theorem 12. *With the same notation as in the Corollary to Theorem 2, we have*

$$\operatorname{Re} B(\chi) + \frac{1}{2} \sum_{n=1}^{\infty} \left(\frac{1}{\rho_n} + \frac{1}{\bar{\rho}_n} \right) = 0. \tag{17}$$

Proof. We take the logarithmic derivative of the two sides of (16) and apply (14):

$$B(\chi) = \frac{\xi'(0, \chi)}{\xi(0, \chi)} = -\frac{\xi'(1, \bar{\chi})}{\xi(1, \bar{\chi})} = -\sum_{n=1}^{\infty} \left(\frac{1}{1 - \bar{\rho}_n} + \frac{1}{\bar{\rho}_n} \right) - B(\bar{\chi}).$$

Since $L(\rho_n, \chi) = L(1 - \rho_n, \bar{\chi}) = L(\bar{\rho}_n, \bar{\chi}) = L(1 - \bar{\rho}_n, \chi) = 0$, then ρ_n and $1 - \bar{\rho}_n$ are zeros of $L(s, \chi)$. The Lemma follows from this. □

§5. Simplest Theorems on the Zeros

We shall prove several simple statements about the non-trivial zeros that are analogous to the corresponding statements about the zeros of the zeta-function.

Theorem 3. *Let $\rho_n = \beta_n + i\gamma_n$, $n = 1, 2, \ldots$ be the non-trivial zeros of $L(s, \chi)$, where χ is a primitive character modulo k. Let $T \geq 2$. Then*

$$\sum_{n=1}^{\infty} \frac{1}{1 + (T - \gamma_n)^2} \leq c \log kT.$$

Proof. For $s = 2 + iT$ we have (with $\delta = 0$ or 1)

$$\left| \sum_{n=1}^{\infty} \left(\frac{1}{s + \delta + 2n} - \frac{1}{2n} \right) \right| \le \sum_{n \le T} \frac{1}{n} + \sum_{n > T} \frac{|s|}{n^2} \le c_1 \log T.$$

From (16) and (14) we obtain

$$- \operatorname{Re} \frac{L'(s, \chi)}{L(s, \chi)} = \frac{1}{2} \log \frac{k}{\pi} - \operatorname{Re} B(\chi) - \operatorname{Re} \sum_{n=1}^{\infty} \left(\frac{1}{s - \rho_n} + \frac{1}{\rho_n} \right) - \frac{\gamma}{2}$$

$$- \operatorname{Re} \frac{1}{s + \delta} - \operatorname{Re} \sum_{n=1}^{\infty} \left(\frac{1}{s + \delta + 2n} - \frac{1}{2n} \right);$$

From this and from (17) we get

$$\operatorname{Re} \sum_{n=1}^{\infty} \frac{1}{s - \rho_n} \le \sum_{n=1}^{\infty} \frac{\Lambda(n)}{n^2} + c_1 \log kT < c_2 \log kT.$$

Moreover,

$$\operatorname{Re} \frac{1}{s - \rho_n} = \operatorname{Re} \frac{1}{(2 - \beta_n) + i(T - \gamma_n)}$$

$$= \frac{2 - \beta_n}{(2 - \beta_n)^2 + (T - \gamma_n)^2} \ge \frac{1}{4 + (T - \gamma_n)^2}.$$

From these estimates together with the first equation, we obtain the Theorem.

$\square$

Under the assumption and notation of Theorem 3 we have the following corollaries.

Corollary 1. The number of zeros ρ_n for which $T \le |\operatorname{Im} \rho_n| \le T + 1$ does not exceed $c \log kT$.

Corollary 2. We have the estimate

$$\sum_{|T - \gamma_n| > 1} \frac{1}{(T - \gamma_n)^2} \le c_1 \log kT.$$

Theorem 4. *For* $-1 \le \sigma \le 2$, $s = \sigma + it$, $|t| \ge 2$, *we have the identity*

$$\frac{L'(s, \chi)}{L'(s, \chi)} = \sum_{|t - \gamma_n| \le 1} \frac{1}{s - \rho_n} + O(\log k|t|),$$

where χ is a primitive character and the sum runs over all zeros ρ_n of the function $L(s, \chi)$ for which $|t - \operatorname{Im} \rho_n| = |t - \gamma_n| \le 1$.

Proof. Just as in the proof of Theorem 3, we have

$$\frac{L'(s, \chi)}{L'(s, \chi)} = \sum_{n=1}^{\infty} \left(\frac{1}{s - \rho_n} + \frac{1}{\rho_n} \right) + B(\chi) - \frac{1}{2} \log \frac{k}{\pi} + \frac{\gamma}{2} + \frac{1}{s + \delta} + O(\log |t|),$$

where $s = \sigma + it$, $|t| \ge 2$, and $-1 \le \sigma \le 2$.

We subtract from this identity the same identity with $s = 2 + it$, and obtain

$$\frac{L'(s, \chi)}{L(s, \chi)} = \sum_{n=1}^{\infty} \left(\frac{1}{s - \rho_n} - \frac{1}{2 + it - \rho_n} \right) + O(\log|t|).$$

If $|\gamma_n - t| > 1$, then

$$\left| \frac{1}{\sigma + it - \rho_n} - \frac{1}{2 + it - \rho_n} \right| \leq \frac{2 - \sigma}{(\gamma_n - t)^2} \leq \frac{3}{(\gamma_n - t)^2},$$

and the Theorem now follows from Corollary 2.   □

Exercises

1. Let $f(t) = \alpha_0 + \alpha_1 t + \alpha_2 t^2 + \ldots$, where $\alpha_0, \alpha_1, \alpha_2, \ldots$ are arbitrary complex numbers. Prove that for any $n \geq 0$ one can find two polynomials

$$P = p_0 + p_1 t + \ldots + p_n t^n, \quad Q = q_0 + q_1 t + \ldots + q_n t^n$$

such that

$$Qf(t) - P = r_{2n+1} t^{2n+1} + r_{2n+1} t^{2n+2} + \ldots$$

2. Let $p \geq 3$ be a prime number, and let $a, b, c, \ldots, a_i, b_i, i = 0, 1, \ldots$ be integers. We shall say that
(a) the polynomial $F(x) = a_0 + a_1 x + \ldots + a_n x^n + a_{n+1} x^{n+1} + \ldots + a_m x^m$ has degree $n \geq 0$ modulo p if $a_m \equiv \ldots \equiv a_{n+1} \equiv 0 \pmod{p}$, $a_n \not\equiv 0 \pmod{p}$;
(b) the polynomial $G(x) = b_0 + b_1 x + \ldots + b_m x^m$ is congruent to $F(x)$ modulo p if $b_i \equiv a_i \pmod{p}$ for $i = 0, 1, \ldots, m$; and
(c) the number a is a root of $F(x)$ modulo p with multiplicity $k \geq 1$ if $F(x) \equiv (x - a)^k (bx^r + cx^{r-1} + \ldots + d) \pmod{p}$.

(1) Prove that if $F(x)$ is a polynomial of degree n modulo p, and if $a_1, \ldots, a_r$ are distinct modulo p and are roots of $F(x)$ modulo p with multiplicities $k_1, \ldots, k_r$, respectively, then $k_1 + \ldots + k_r \leq n$.
(2) Prove that if $F(a) \equiv 0 \pmod{p}$, then a is a root of $F(x)$ modulo p.
(3) Prove that if $k \geq 1$ and

$$F(a) \equiv \frac{1}{1!} F'(a) \equiv \ldots \equiv \frac{1}{(k-1)!} F^{(k-1)}(a) \equiv 0 \pmod{p},$$

then a is a root of $F(x)$ modulo p of multiplicity k.
3. Let

$$f(x) = x^3 + ax + b, \quad F(x) = \pm(f(x))^{(p-1)/2} + 1,$$

$$g(x) = 2f(x)(\pm(f(x))^{(p-1)/2} + 1) + f'(x)(x^p - x).$$

Prove that each root of $F(x)$ modulo p is at least a double root of $g(x)$.

Show that this implies that the number N_p of solutions of the congruence $y^2 \equiv x^3 + ax + b \,(\mathrm{mod}\, p)$ satisfies

$$|N_p - p| \le (p + 3)/2.$$

4. Let n be an odd positive integer,

$$f(x) = ax^n + bx^{n-1} + \ldots + c, \quad (a, p) = 1,$$

$$F_1(x) = f^{(p-1)/2}(x) + 1, \quad F_2(x) = f^{(p-1)/2}(x) - 1, \quad F(x) = F_i(x).$$

Decomposing the difference $F(x^p) - F(x)$ into powers of $H = x^p - x$ and applying exercise 1, prove that there exists a polynomial $g(x)$ of degree m with

$$1 \le m \le kp + (k^2 + k)(n - 1) + \frac{p-1}{2}\, n$$

such that each root of $F(x)$ is a root of g of multiplicity $2k + 1$, where $k \le (p - 1)/4$.

5. Under the conditions of exercise 4, prove that

$$\left| \sum_{x=1}^{p} \left(\frac{f(x)}{p} \right) \right| < 2n\sqrt{p}.$$

6. Prove that if $V(X)$ and $N(X)$ are the numbers of residues and non-residues, respectively, modulo p in the interval $[1, X]$, then

$$V(X) = \frac{1}{2} X + \theta \sqrt{p} \log p, \quad N(X) = \frac{1}{2} X - \theta \sqrt{p} \log p, \quad |\theta| \le 1.$$

7. Let $X = X(p) \to +\infty$ as $p \to +\infty$ and let $V(Y) = Y/2 + o(Y)$ for each $Y \ge X$. Denote by $n = n(p)$ the smallest quadratic non-residue modulo p. Show that

$$n \le cX^{1/\sqrt{e}}.$$

8. Prove that if $k \ge 1$ and $1 \le Z < p$, then

$$\sum_{\lambda=1}^{p} \left(\sum_{m=1}^{Z} \left(\frac{\lambda + m}{p} \right) \right)^{2k} \le (2k)^k Z^k p + 4kZ^{2k} \sqrt{p}.$$

9. Let U and V be integers,

$$p \ge p_0(\varepsilon), \ p^{\theta,\,5+\varepsilon} \le U < p, \ p^\varepsilon < V < p, \ W = \sum_{u}^{U} \sum_{v}^{V} \left(\frac{u + v}{p} \right),$$

where u and v in the last sum run over U and V, respectively, different values modulo p. Then

$$|W| \le cUVp^{-\delta}, \quad \delta = \delta(\varepsilon) > 0, \quad c = c(\varepsilon).$$

10. Let $X \ge p^{0,25+\varepsilon}$. Then

$$|S| = \left| \sum_{m \le X} \left(\frac{m}{p} \right) \right| < cXp^{-\delta}, \quad \delta = \delta(\varepsilon) > 0, \quad c = c(\varepsilon).$$

11. Let $n = n(p)$ be the smallest quadratic non-residue modulo p. Prove that

$$n = n(p) = O(p^{\frac{1}{4\sqrt{e}}+\varepsilon}).$$

12. Let x_r be a sequence of real numbers such that $\| x_r - x_m \| > \delta > 0, r \neq m$. Then for any sequence a_n

$$\sum_{r=1}^{R} \left| \sum_{n=M+1}^{M+N} a_n e^{2\pi i n x_r} \right|^2 \leq c \left(N + \frac{1}{\delta} \right) \sum_{n=M+1}^{M+N} |a_n|^2.$$

13. For $\pi(x)(1 + o(1))$ values of $p, p \leq x$, the smallest quadratic non-residue $n = n(p)$ and the smallest prime quadratic residue $v = v(p)$ do not exceed $c \log^{2+\varepsilon} x$.

Chapter IX. Prime Numbers in Arithmetic Progressions

The method of complex integration together with the results of L-functions proved in Chapter VIII will enable us to write down an explicit formula connecting the sum of values of the function $\Lambda(n)$ over the integers lying in a given arithmetic progression with the zeros of an L-function. This explicit formula together with a theorem on the boundary of the zeros of the L-function will yield the prime number theorem for arithmetic progressions. We shall always assume below that $k \leq x$.

§1. An Explicit Formula

We introduce two functions analogous to the Chebyshev ψ-function.

Definition. Let χ be an arbitrary character modulo k. The functions $\psi(x, \chi)$ and $\psi(x; k, l)$ for $1 \leq l \leq k$ and $(l, k) = 1$ are defined by the equations

$$\psi(x, \chi) = \sum_{n \leq x} \Lambda(n)\chi(n),$$

$$\psi(x; k, l) = \sum_{\substack{n \equiv 1 \pmod k \\ n \leq x}} \Lambda(n).$$

Using the orthogonality properties of characters (Chapter VIII, property 4), we have

$$\psi(x; k, l) = \frac{1}{\varphi(k)} \sum_{\chi \bmod k} \psi(x, \chi)\bar{\chi}(l)$$

$$= \frac{1}{\varphi(k)} \sum_{\substack{n \leq x \\ (n, k) = 1}} \Lambda(n) + \frac{1}{\varphi(k)} \sum_{\chi \neq \chi_0} \psi(x, \chi)\bar{\chi}(l). \tag{1}$$

The first summand differs from $(1/\varphi(k)) \psi(x)$ by the quantity

$$\frac{1}{\varphi(k)} \sum_{\substack{n \leq x \\ (n, k) > 1}} \Lambda(n) \leq \frac{\ln^2 x}{\varphi(k)}.$$

Furthermore, if $k_1 | k$ and if χ_1 is the primitive character modulo k_1 that induces the character χ, then

$$\psi(x, \chi) = \psi(x, \chi_1) + \theta \sum_{\substack{n \le x \\ (n, k) > 1}} \Lambda(n) = \psi(x, \chi_1) + \theta_1 \log^2 x,$$

$$|\theta| \le 1, \quad |\theta_1| \le 1.$$

Thus, the study of $\psi(x; k, 1)$ is reduced to the study of $\psi(x, \chi_1)$, where χ_1 is a primitive character modulo k_1, and $k_1 | k$.

Theorem 1. *Let χ be a primitive character modulo k, and let $2 \le T \le x$. Then*

$$\psi(x, \chi) - \psi(k, \chi) = - \sum_{|\operatorname{Im}\rho| \le T} \frac{x^\rho - k^\rho}{\rho} + O\left(\frac{x \log^2 x}{T}\right),$$

where ρ runs through the non-trivial zeros of $L(s, \chi)$.

Proof. By Corollary 1 of Chapter VIII, we can find T_1 with $T \le T_1 \le T + 1$ such that $|T_1 - \operatorname{Im}\rho_n| > 1/c \log kT$, where the ρ_n are the zeros of $L(s, \chi)$. We consider the rectangle Γ with vertices at the points $b - iT_1$, $b + iT_1$, $-0.5 + iT_1$, $-0.5 - iT_1$, where $b = 1 + (\log x)^{-1}$.

Integrating $(x^s/s)(d/ds)(\ln L(s, \chi))$ around the contour Γ, we find that

$$\frac{1}{2\pi i} \int_\Gamma \left\{ -\frac{L'(s, \chi)}{L(s, \chi)} \right\} \frac{x^s - k^s}{s} = - \sum_{|\operatorname{Im}\rho| \le T_1} \frac{x^\rho - k^\rho}{\rho} + \theta \ln x,$$

where $|\theta| \le 1$. By Theorem 1 of Chapter V with $\alpha = 1$ and $x = N + 0.5$,

$$\psi(x, \chi) = \frac{1}{2\pi i} \int_{b - iT_1}^{b + iT_1} \left\{ -\frac{L'(s, \chi)}{L(s, \chi)} \right\} \frac{x^s}{s} ds + O\left(\frac{x \log^2 x}{T}\right).$$

It remains to estimate the integrals on the top, the bottom, and the left side of the rectangle Γ. The integrals on the top and bottom sides of Γ can be estimated in the same way. Using Theorem 4 of Chapter VIII and the chosen T_1, we obtain

$$\left| \frac{1}{2\pi i} \int_{-0.5 + iT_1}^{b + iT_1} \left\{ -\frac{L'(s, \chi)}{L(s, \chi)} \right\} \frac{x^s}{s} ds \right| \le \frac{e}{2\pi} \int_{-0.5}^{b} \left| \frac{L'(\sigma + iT_1)}{L(\sigma + iT_1)} \right| \frac{x}{T_1} d\sigma$$

$$= \frac{e}{2\pi} \int_{-0.5}^{b} \left| \sum_{|T_1 - \gamma_n| \le 1} \frac{1}{\sigma - \sigma_n + i(T_1 - \gamma_n)} + O(\log kT) \right| \frac{x}{T_1} d\sigma$$

$$= O\left(\frac{x \log^2 kT}{T}\right) = O\left(\frac{x \log^2 x}{T}\right).$$

The integral on the left side of Γ is estimated thus:

$$\left| \frac{1}{2\pi} \int_{-T_1}^{+T_1} \left\{ -\frac{L'(-0,5 + it)}{L(-0,5 + it)} \right\} \frac{x^{-0,5 + it}}{-0,5 + it} dt \right|$$

$$\le x^{-0.5} \int_{-T_1}^{+T_1} \left| \frac{L'(-0,5 + it)}{L(-0,5 + it)} \right| \frac{dt}{0,5 + |t|} = O\left(\frac{\log^2 kT}{\sqrt{x}}\right),$$

since by Theorem 4 of Chapter VIII

$$\left| \frac{L'(-0,5+it)}{L(-0,5+it)} \right| = O(\ln k(|t|+2)).$$

Combining these estimates, we obtain the Theorem.   □

§2. Theorems on the Boundary of the Zeros

Just as in the proof of Theorem 5 in Chapter IV on the boundary for the zeros of the zeta-function, we shall need the inequality

$$3 + 4\cos\varphi + \cos 2\varphi \geq 0,$$

where φ is a real number. We shall also use the upper estimates for the quantity

$$- \operatorname{Re} \frac{L'(s, \chi)}{L(s, \chi)}$$

where $s = \sigma + it$, $\chi = \chi_1$, and $s = \sigma + i2t$,

The boundary for the real zeros of $L(s, \chi)$ for real characters will present a special difficulty.

Theorem 2. *If χ is a complex character modulo k and $s = \sigma + it$, then $L(s, \chi)$ has no zeros in the domain*

$$\operatorname{Re} s = \sigma \geq 1 - \frac{c}{\log k(|t|+2)}.$$

If χ is a real character modulo k and $s = \sigma + it$, then $L(s, \chi)$ has no zeros in the domain

$$\operatorname{Re} s = \sigma \geq 1 - \frac{c}{\log k(|t|+2)}, \quad |t| > 0.$$

Proof. We first consider the case of primitive characters. Let χ be a complex character, $s = \sigma + it$, $\sigma > 1$, $t \geq 0$. Then $\chi(n) = e^{i\omega(n)}$ and

$$- \frac{L'(s, \chi)}{L(s, \chi)} = \sum_{n=1}^{\infty} \frac{\Lambda(n)\chi(n)}{n^s} = \sum_{n=1}^{\infty} \Lambda(n) n^{-\sigma} e^{-it\log n + i\omega(n)};$$

$$- \operatorname{Re} \frac{L'(s, \chi)}{L(s, \chi)} = \sum_{n=1}^{\infty} \Lambda(n) n^{-\sigma} \cos\{t\log n - \omega(n)\},$$

$$- \operatorname{Re} \frac{L'(\sigma + i2t, \chi^2)}{L(\sigma + i2t, \chi^2)} = \sum_{n=1}^{\infty} \Lambda(n) n^{-\sigma} \cos 2\{t\log n - \omega(n)\}.$$

Therefore,

$$3\left\{-\frac{L'(\sigma, \chi_0)}{L(\sigma, \chi_0)}\right\} + 4\left\{-\operatorname{Re}\frac{L'(\sigma + it, \chi)}{L(\sigma + it, \chi)}\right\}$$

$$+ \left\{-\operatorname{Re}\frac{L'(\sigma + i2t, \chi^2)}{L(\sigma + i2t, \chi^2)}\right\} \geq 0. \tag{2}$$

We shall estimate each term in (2) from above. For this we apply Lemma 12 of Chapter VIII and the estimate for $|\Gamma'(s)/\Gamma(s)|$ (cf., for example, the proof of Theorem 3 in Chapter VIII):

$$-\frac{L'(\sigma, \chi_0)}{L(\sigma, \chi_0)} = \sum_{n=1}^{\infty} \chi_0(n)\Lambda(n)n^{-\sigma} \leq -\frac{\zeta'(\sigma)}{\zeta(\sigma)} < \frac{1}{\sigma - 1} + c_1;$$

$$-\operatorname{Re}\frac{L'(s, \chi)}{L(s, \chi)} = -\operatorname{Re}\frac{L(\sigma + it, \chi)}{L(\sigma + it, \chi)}$$

$$= \frac{1}{2}\log\frac{k}{\pi} - \operatorname{Re}\sum_{n=1}^{\infty}\left(\frac{1}{s - \rho_n} + \frac{1}{\rho_n}\right) - \frac{\gamma}{2} - \operatorname{Re}\frac{1}{s + \delta}$$

$$- \operatorname{Re}B(\chi) - \operatorname{Re}\sum_{n=1}^{\infty}\left(\frac{1}{s + \delta + 2n} - \frac{1}{2n}\right) \leq c_2\log k(t + 2)$$

$$- \operatorname{Re}\sum_{n=1}^{\infty}\frac{1}{s - \rho_n}. \tag{3}$$

If χ_1 denotes the primitive character that induces χ^2, then $\chi_1 \neq \chi$ and

$$\left|\frac{L'(s, \chi^2)}{L(s, \chi^2)} - \frac{L'(s, \chi_1)}{L(s, \chi_1)}\right| \leq \sum_{p\mid k}\frac{p^{-\sigma}\log p}{1 - p^{-\sigma}} \leq \sum_{p\mid k}\log p \leq \log k.$$

Consequently, applying the estimate (3), we find that

$$-\operatorname{Re}\frac{L'(\sigma + i2t, \chi^2)}{L(\sigma + i2t, \chi^2)} \leq -\operatorname{Re}\frac{L'(\sigma + i2t, \chi_1)}{L(\sigma + i2t, \chi_1)} + \log k \leq c_3\log k(t + 2). \tag{4}$$

Since

$$\operatorname{Re}\frac{1}{s - \rho} = \frac{\sigma - \beta}{|s - \rho|^2} \geq 0,$$

then

$$\frac{3}{\sigma - 1} - 4\operatorname{Re}\frac{1}{s - \rho} + c\log k(t + 2) \geq 0. \tag{5}$$

Now let $\rho = \beta + i\gamma$ be a zero of $L(s, \chi)$. Without loss of generality, we can assume that $\gamma \geq 0$. In (5) we set $t = \gamma$, $\sigma = 1 + 1/2c \log k(t + 2)$. This gives

$$\beta \leq 1 - 1/14c\log k(\gamma + 2).$$

We shall now prove the assertion of the Theorem for real primitive characters χ. We have $\chi^2 = \chi_0$,

$$\left| \frac{L'(s, \chi^2)}{L(s, \chi^2)} - \frac{\zeta'(s)}{\zeta(s)} \right| \leq \log k.$$

From Theorems 3 and 4 of Chapter IV,

$$- \operatorname{Re} \frac{\zeta'(s)}{\zeta(s)} < \operatorname{Re} \frac{1}{s-1} + c_2 \log(t + 2).$$

We insert this estimate and estimate (3) into (2), and we let $t = \gamma$, where $\rho = \beta + i\gamma$ is a zero of $L(s, \chi)$, $\gamma \geq 0$. This gives

$$\frac{3}{\sigma - 1} - \frac{4}{\sigma - \beta} + \operatorname{Re} \frac{1}{\sigma - 1 + i2\gamma} + c_4 \log k(\gamma + 2) \geq 0;$$

$$\operatorname{Re} \frac{1}{\sigma - 1 + i2\gamma} = \frac{\sigma - 1}{(\sigma - 1)^2 + 4\gamma^2};$$

$$\frac{4}{\sigma - \beta} \leq \frac{3}{\sigma - 1} + \frac{\sigma - 1}{(\sigma - 1)^2 + 4\gamma^2} + c_4 \log k(\gamma + 2).$$

We consider two cases: The case of large γ and the case of small γ. Let $\gamma > \chi/\log k$, where $0 < \chi < 1/5\, c_4$, and χ is an absolute constant. Setting $\sigma = 1 + \kappa/\log k\, (\gamma + 2)$, we find

$$\beta \leq 1 - \frac{c_5}{\log k(\gamma + 2)}, \quad c_5 \geq \frac{3}{5c_4 + 16\kappa^{-1}}.$$

Now let $0 < \gamma < \kappa/\log k$. Then, applying (3), we have

$$- \frac{L'(\sigma, \chi)}{L(\sigma, \chi)} < c_2 \log k - \sum_{n=1}^{\infty} \frac{1}{\sigma - \rho_n} < c_2 \log k - \frac{2(\sigma - \beta)}{(\sigma - \beta)^2 + \gamma^2}, \qquad (6)$$

since the zeros ρ of the function $L(s, \chi)$ have the form $\rho = \beta \pm i\gamma$. Furthermore,

$$- \frac{L'(\sigma, \chi)}{L(\sigma, \chi)} = \sum_{n=1}^{\infty} \chi(n) \Lambda(n) n^{-\sigma} \geq - \sum_{n=1}^{\infty} \Lambda(n) n^{-\sigma}$$

$$= \frac{\zeta'(\sigma)}{\zeta(\sigma)} > - \frac{1}{\sigma - 1} - c_6.$$

From this and from (6)

$$\frac{2(\sigma - \beta)}{(\sigma - \beta)^2 + \gamma^2} < \frac{1}{\sigma - 1} + c_r \log k.$$

Let $\sigma = 1 - \lambda/\log k$ and $\kappa = \lambda/10$. Then the last inequality implies that

$$\beta \leq 1 - \frac{\lambda}{10 \log k}, \quad \lambda > \frac{2}{3c_7}.$$

This proves the Theorem for primitive characters χ. For an arbitrary character χ, the Theorem follows from what has already been proved together with Lemma 8 of Chapter VIII. $\qquad\qquad\qquad\qquad\qquad\qquad\qquad\qquad\qquad\qquad\qquad$ $\square$

We next turn to the study of the distribution of the real zeros of $L(s, \chi)$ for a real primitive character χ. The boundary for such zeros is, at the present time, much more crude than that which has been obtained in Theorem 2. First we must estimate $L(1, \chi)$ from below.

Lemma 1. *Let χ be a real primitive character modulo k. Then*

$$L(1, \chi) \geq \frac{c}{\sqrt{k} \log^2 k}.$$

Proof. Using the estimate for character sums (Lemma 5 of Chapter VIII) together with partial summation (Lemma C of Chapter I), we find for $m > 0$ that

$$\left| \sum_{m < n \leq M} \frac{\chi(n)}{n} \right| = \left| \int_m^M \left(\sum_{m < n \leq x} \chi(n) \right) x^{-2} \, dx + \frac{1}{M} \sum_{m < n \leq M} \chi(n) \right| \leq c_1 \frac{\sqrt{k} \log k}{m};$$

$$\left| L(1, \chi) - \sum_{n \leq x} \frac{\chi(n)}{n} \right| \leq \frac{c_1 \sqrt{k} \log k}{x}.$$

For $1/2 \leq t < 1$, we consider the function $H(t)$ defined by

$$H(t) = \sum_{n=1}^{\infty} a_n t^n, \quad a_n = \sum_{d \mid n} \chi(d).$$

If $n = p_1^{\alpha_1} \ldots p_u^{\alpha_u}$ is the canonical decomposition of n into prime factors, then

$$a_n = \prod_{r=1}^{u} (1 + \chi(p_r) + \ldots + \chi(p_r^{\alpha_r})).$$

Therefore, $a_n \geq 0$ and $a_{m^2} \geq 1$. This implies that

$$H(t) > \sum_{m=1}^{\infty} t^{m^2} > \int_2^{\infty} t^{u^2} \, du > \int_0^{\infty} t^{u^2} \, du - 2$$

$$= \frac{\sqrt{\pi}}{2\sqrt{-\ln(1 - (1 - t))}} - 2 > \frac{1}{2\sqrt{1 - t}}.$$

Moreover,

$$H(t) = \sum_{m=1}^{\infty} \left(\sum_{n \mid m} \chi(n) \right) t^m = \sum_{n=1}^{\infty} \chi(n) \sum_{n=1}^{\infty} t^{rn} = \sum_{n=1}^{\infty} \frac{\chi(n) t^n}{1 - t^n}.$$

We estimate the difference $H(t) - \dfrac{L(1, \chi)}{1 - t} = G(t)$ from above:

$$G(t) = \sum_{n=1}^{\infty} \chi(n) \frac{t^n}{1 - t^n} - \sum_{n=1}^{\infty} \frac{\chi(n)}{n} \frac{t^n}{1 - t} + \sum_{n=1}^{\infty} (S_n - S_{n+1}) \frac{t^n}{1 - t} - \frac{L(1, \chi)}{1 - t}$$

$$= \sum_{n=1}^{\infty} \chi(n) \left\{ \frac{t^n}{1 - t^n} - \frac{t^n}{n(1 - t)} \right\} - \sum_{n=0}^{\infty} S_{n+1} t^n,$$

where $S_n = \sum_{m=n}^{\infty} \dfrac{\chi(m)}{m}$. We have

$$|S_n| \leq \frac{c_1 \sqrt{k} \log k}{n},$$

$$\left| \sum_{n=0}^{\infty} S_{n+1} t^n \right| < c_1 \sqrt{k} \log k \sum_{n=1}^{\infty} \frac{t^n}{n} = c_1 \sqrt{k} \log k \log \frac{1}{1 - t};$$

$$\left| \sum_{n=1}^{\infty} \chi(n) \left\{ \frac{t^n}{1 - t^n} - \frac{t^n}{n(1 - t)} \right\} \right| = \left| \sum_{n=1}^{\infty} \left(\sum_{m=1}^{n} \chi(m) \right) \left\{ \frac{t^n}{1 - t^n} - \frac{t^n}{n(1 - t)} \right. \right.$$

$$\left. \left. - \frac{t^{n+1}}{1 - t^{n+1}} + \frac{t^{n+1}}{(n + 1)(1 - t)} \right\} \right| \leq \frac{c_1 \sqrt{k} \log k}{1 - t} \sum_{n=1}^{\infty} \left| \frac{t^n}{1 + t + \ldots + t^{n-1}} \right.$$

$$\left. - \frac{t^{n+1}}{1 + t + \ldots + t^n} - \frac{t^n}{n(n + 1)} - \frac{(1 - t)t^n}{n + 1} \right|$$

$$< \frac{c_1 \sqrt{k} \log k}{1 - t} \sum_{n=1}^{\infty} \left(\frac{t^n}{1 + t + \ldots + t^{n-1}} - \frac{t^{n+1}}{1 + t + \ldots + t^n} \right.$$

$$\left. - \frac{t^n}{n(n + 1)} \right) + c_1 \sqrt{k} \log k \sum_{n=1}^{\infty} \frac{t^n}{n + 1} < 2 c_1 \sqrt{k} \log k \log \frac{1}{1 - t}.$$

We obtain

$$|G(t)| < 3 c_1 \sqrt{k} \log k \log \frac{1}{1 - t}.$$

From this we get

$$\frac{L(1, \chi)}{1 - t} = H(t) - G(t) > \frac{1}{2\sqrt{1 - t}} + 3 c_1 \sqrt{k} \log k \log(1 - t) - 2.$$

Let

$$t = 1 - \frac{1}{c_0 k \log^4 k}, \qquad c_0 = (64(c_1 + 1))^2.$$

Then

$$\frac{L(1, \chi)}{1 - t} > \frac{1}{4}\sqrt{c_0 k}\log^2 k.$$

This is what we had to prove.   □

We shall next prove the theorem of A. Page on the boundary of the real zeros, and derive some consequences from it.

Theorem 3. *Let χ be a real primitive character modulo k. Then*

$$L(\sigma, \chi) \neq 0 \quad \text{for } \sigma > 1 - \frac{c}{\sqrt{k}\log^4 k}.$$

Proof. We shall consider σ in the interval $[1 - (1/8)\log k, 1]$. By the mean value theorem

$$L(1, \chi) = L(\sigma, \chi) + (1 - \sigma)L'(\sigma_1, \chi),$$

where $\sigma \leq \sigma_1 \leq 1$. Applying partial summation (Lemma C, Chapter I) and the estimate for character sums (Lemma 5, Chapter VIII), we find that

$$|L'(\sigma_1, \chi)| = \left| \sum_{n=1}^{\infty} \frac{\chi(n)\log n}{n^{\sigma_1}} \right|$$

$$\leq \sum_{n \leq k} \frac{\log n}{n^{\sigma_1}} + \int_k^{\infty} \left| \sum_{k < n \leq x} \chi(n) \right| \left(\frac{1}{x^{1+\sigma_1}} + \frac{\log x}{x^{1+\sigma_1}} \right) dx \leq c_1 \log^2 k.$$

From this and Lemma 1, we obtain

$$L(\sigma, \chi) \geq L(1, \chi) - (1 - \sigma)c_1 \log^2 k$$

$$> \frac{c_0}{\sqrt{k}\log^2 k} - (1 - \sigma)c_1 \log^2 k > 0,$$

if

$$\sigma > 1 - \frac{c}{\sqrt{k}\log^4 k}, \quad c < \frac{c_0}{c_1}.$$

This proves the Theorem.   □

As a consequence of Theorems 2 and 3 and of Chapter IV, Theorem 5, we find that $L(1, \chi)$ is non-zero for any χ, i.e. different from a zero of $\xi(0, \chi)$ for any χ (cf. the proof of Theorem 2 of Chapter VIII).

In applications, the boundary for the real zeros of $L(s, \chi)$ obtained in Theorem 3 is often not sufficiently strong. However, the moduli for which there might be a large real zero occur extremely rarely. This circumstance makes it possible for us to prove a statement that will allow us to use Theorem 3 successfully in many applications (cf., for example, Chapter X).

Theorem 4. *Let χ_1 be a real primitive character modulo k_1 and let χ_2 be a real primitive character modulo k_2, where $\chi_1 \not\equiv \chi_2$. Let $L(s, \chi_1)$ and $L(s, \chi_2)$ have real zeros β_1 and β_2, respectively. Then*

$$\min(\beta_1, \beta_2) < 1 - \frac{c}{\log k_1 k_2}.$$

Proof. We consider the character $\chi(n) = \chi_1(n)\chi_2(n)$ modulo $k_1 k_2$ (cf. the Corollary to Lemma 4 of Chapter VIII). Moreover,

$\chi^{(n)} \not\equiv \chi_{.0}(n)$, since $\chi_1 \not\equiv \chi_2$. For $\sigma > 1$ we have

$$0 \leq \sum_{n=1}^{\infty} \Lambda(n)(1 + \chi_1(n))(1 + \chi_2(n))n^{-\sigma}$$

$$= -\frac{\zeta'(\sigma)}{\zeta(\sigma)} - \frac{L'(\sigma, \chi_1)}{L(\sigma, \chi_1)} - \frac{L'(\sigma, \chi_2)}{L(\sigma, \chi_2)} - \frac{L'(\sigma, \chi)}{L(\sigma, \chi)}. \tag{7}$$

Just as in the proof of inequality (4), we find that

$$-\frac{L'(\sigma, \chi)}{L(\sigma, \chi)} < c_1 \log k_1 k_2.$$

Moreover, from (3),

$$-\frac{L'(\sigma, \chi_1)}{L(\sigma, \chi_1)} < c_1 \log k_1 - \frac{1}{\sigma - \beta_1},$$

$$-\frac{L'(\sigma, \chi_2)}{L(\sigma, \chi_2)} < c_1 \log k_2 - \frac{1}{\sigma - \beta_2}.$$

We insert this estimate into (7):

$$\frac{1}{\sigma - \beta_1} + \frac{1}{\sigma - \beta_2} + < \frac{1}{\sigma - 1} + c_2 \log k_1 k_2.$$

Setting $\sigma = 1 + \dfrac{1}{2c_2 \log k_1 k_2}$, we find that

$$\min(\beta_1, \beta_2) \leq 1 - \frac{1}{7c_2 \log k_1 k_2}.$$

This proves the Theorem. □

Corollary 1. *There is at most one character χ modulo k such that the corresponding function $L(s, \chi)$ has a zero β satisfying the condition*

$$\beta \geq 1 - \frac{c}{\log k}.$$

Proof. If χ_1 and χ_2 are two distinct characters modulo k, then they are induced by two distinct characters χ_1^* and χ_2^* modulo k_1^* and k_2^*, respectively, where k_1^* and k_2^* do not exceed k. This proves the Corollary.   □

Corollary 2. *Let $3 \le k \le x$: There exists at most one k_0 with $3 \le k_0 \le x$ and at most one real primitive character χ_1 modulo k_0 for which $L(s, \chi_1)$ has a simple real zero β_1 such that*

$$\beta_1 \ge 1 - \frac{c}{\log x}.$$

Moreover, if χ is a real character modulo k and if $L(s, \chi)$ satisfies

$$L(\beta, \chi) = 0, \quad \beta \ge 1 - \frac{c}{\log x},$$

then $k \equiv 0 \pmod{k_0}$.

Proof. If $m \ge 2$ and if β_1 is a zero of $L(s, \chi_1)$ of multiplicity m, then the same proof as in Theorem 4 shows that

$$0 \le \sum_{n=1}^{\infty} \Lambda(u)(1 + \chi_1(n))n^{-\sigma} = -\frac{\zeta'(\sigma)}{\zeta(\sigma)} - \frac{L'(\sigma, \chi_1)}{L(\sigma, \chi_1)}, \quad \sigma > 1;$$

$$-\frac{\zeta'(\sigma)}{\zeta(\sigma)} < \frac{1}{\sigma - 1} + c_1, \quad -\frac{L'(\sigma, \chi_1)}{L(\sigma, \chi_1)} < c_1 \log x - \frac{m}{\sigma - \beta_1};$$

$$\frac{2}{\sigma - \beta_1} \le \frac{m}{\sigma - \beta_1} < \frac{1}{\sigma - 1} + c_2 \log x; \quad \beta_1 \le 1 - \frac{1}{7 c_2 \log x},$$

which contradicts the hypothesis of the Corollary. Moreover, if there is another real primitive character χ_2 not identically equal to χ_1 for which $L(s, \chi_2)$ has a large real zero β_2,

$$\beta_2 \ge 1 - \frac{c}{\log x},$$

then the Theorem is contradicted. Now let χ be a real character modulo k such that

$$L(\beta, \chi) = 0, \quad \beta \ge 1 - \frac{c}{\log x}.$$

If χ_2 is the primitive character modulo k_2 that induces χ, then $k \equiv 0 \pmod{k_2}$ and

$$L(\beta, \chi_2) = 0, \quad \beta \ge 1 - \frac{c}{\log x}.$$

If follows from this that $k_2 = k_0$ and $\chi_2 \equiv \chi_1$. This completes the proof of the Corollary.   □

Corollary 3. *With the same notation and hypotheses as in Corollary 2, we have*

$$k_0 \geq \frac{c' \log^2 x}{(\log \log x)^8}.$$

Proof. Using Theorem 3 and the hypotheses of the Corollary, we have

$$1 - \frac{c}{\log x} < \beta_1 \leq 1 - \frac{c_1}{\sqrt{k_0} \log^4 k_0}.$$

The Corollary follows from this.   □

The following theorem of C.L. Siegel improves the previous estimate for the size of a real zero.

Theorem 5. *For any $\varepsilon > 0$ there exists a constant $c = c(\varepsilon) > 0$ such that, if χ is a real character modulo k and if β is a real zero of $L(s, \chi)$, then*

$$\beta \leq 1 - \frac{c(\varepsilon)}{k^\varepsilon}.$$

In order to prove Siegel's Theorem, we need some additional results.

Lemma 2. *Let χ_1 and χ_2 be distinct real primitive characters modulo k_1 and k_2, respectively. Let.*

$$F(s) = \zeta(s) L(s, \chi_1) L(s, \chi_2) L(s, \chi_1 \chi_2).$$

Then for $9/10 < \sigma < 1$ we have

$$F(\sigma) > \frac{1}{2} - \frac{c\lambda}{1 - \sigma} (k_1 k_2)^{8(1-\sigma)},$$

where $\lambda = L(1, \chi_1) L(1, \chi_2) L(1, \chi_1 \chi_2)$.

Proof. Note that $\chi_1 \chi_2$ is a non-principal character modulo $k_1 k_2$. Consequently, $F(s)$ is regular in the entire s-plane, except at the point $s = 1$, where it has a pole of order one with residue $\lambda = L(1, \chi_1) L(1, \chi_2) L(1, \chi_1 \chi_2)$. For Re $s > 1$ we decompose $F(s)$ into a Dirichlet series:

$$F(s) = \sum_{n=1}^{\infty} b_n n^{-3}.$$

Since Re $s > 1$,

$$F(s) = \prod_p \left(1 - \frac{1}{p^s}\right)^{-1} \left(1 - \frac{\chi_1(p)}{p^s}\right)^{-1} \left(1 - \frac{\chi_2(p)}{p^s}\right)^{-1} \left(1 - \frac{\chi_1(p)\chi_2(p)}{p^s}\right)^{-1},$$

and

$$\chi_1(p) = 0, \pm 1, \chi_2(p) = 0, \pm 1,$$

then it is easy to see that

$$b_1 = 1 \quad \text{and} \quad b_n \geq 1 \text{ for } n > 1.$$

Indeed, if $\chi_1(p) = -1$ and $\chi_2(p) = +1$, then

$$\prod_1 = \prod_p{}' \left(1 - \frac{1}{p^s}\right)^{-1} \left(1 + \frac{1}{p^s}\right)^{-1} \left(1 - \frac{1}{p^s}\right)^{-1} \left(1 + \frac{1}{p^s}\right)^{-1}$$

$$= \prod_p{}' \left(1 - \frac{1}{p^{2s}}\right)^{-2} = \left(\prod_p{}' \left(1 + \frac{1}{p^{2s}} + \frac{1}{p^{1s}} + \cdots \right)\right)^2$$

If $\chi_1(p) = 0$, $\chi_2(p) = +1$, then

$$\prod_2 = \prod_p{}'' \left(1 - \frac{1}{p^s}\right)^{-1} \left(1 - \frac{1}{p^s}\right)^{-1} = \left(\prod_p{}'' \left(1 + \frac{1}{p^s} + \frac{1}{p^{2s}} + \cdots \right)\right)^2.$$

If $\chi_1(p) = 0$, $\chi_2(p) = -1$, then

$$\prod_3 = \prod_p{}''' \left(1 - \frac{1}{p^s}\right)^{-1} \left(1 + \frac{1}{p^s}\right)^{-1} = \prod_p{}''' \left(1 + \frac{1}{p^{2s}} + \frac{1}{p^{4s}} + \cdots \right)$$

If $\chi_1(p) = \chi_2(p) = 0$, then

$$\prod_4 = \prod_p{}'''' \left(1 - \frac{1}{p^s}\right)^{-1} = \prod_p{}'''' \left(1 + \frac{1}{p^s} + \frac{1}{p^{2s}} + \cdots \right).$$

The remaining possible cases are similar to the ones already considered. Multiplying out all the products $\prod_i$, we obtain a Dirichlet series $F(s)$ with $b_1 = 1$ and $b_n \geq 0$.

Consequently, for $|s - 2| < 1$

$$F(s) = \sum_{m=0}^{\infty} a_m(2 - s)^m, \quad a_0 \geq 1, \quad a_m \geq 0,$$

since

$$F^{(m)}(2) = (-1)^m \sum_{n=1}^{\infty} \frac{b_n}{n^2} \log^m n.$$

The function $g(s)$ defined by the equation

$$g(s) = F(s) - \frac{\lambda}{s-1},$$

is regular in the entire s-plane. Therefore, the identity

$$g(s) = F(s) - \frac{\lambda}{s-1} = \sum_{m=0}^{\infty} (a_m - \lambda)(2 - s)^m \tag{8}$$

holds in the circle

$$|s - 2| \leq 3/2. \tag{9}$$

We estimate $g(s)$ in the circle (9). On the boundary $|s - 2| = 3/2$, we have

$$\zeta(s) = O(1), \quad \frac{1}{s - 1} = O(1);$$

$$|L(s, \chi_1)| < ck_1, \quad |L(s, \chi_2)| < ck_2, \quad |L(s, \chi_1\chi_2)| < ck_1 k_2$$

(by the Corollary to Lemma 9, Chapter VIII). Consequently,

$$|g(s)| < e_1(k_1 k_2)^2, \quad |s - 2| = 3/2.$$

By the maximum principle, the last inequality also holds inside the circle (9). Estimating $a_m - \lambda$ in (8) by Cauchy's theorem on the coefficients of a power series, we find that

$$|a_m - \lambda| < c_2(k_1 k_2)^2 (2/3)^m, \quad m = 0, 1, 2, \ldots$$

For $M > 1$ and $9/10 < \sigma < 1$, we have

$$\sum_{m=M}^{\infty} |a_m - \lambda|(2 - \sigma)^m \leq \sum_{m=M}^{\infty} c_2(k_1 \ k_2)^2 \left(\frac{2}{3}(2 - \sigma) \right)^m < c_3(k_1 k_2)^2 \left(\frac{11}{15} \right)^M;$$

$$F(\sigma) - \frac{\lambda}{\sigma - 1} \geq 1 - \lambda \sum_{m=0}^{M-1} (2 - \sigma)^m - c_3(k_1 k_2)^2 \left(\frac{11}{15} \right)^M$$

$$= 1 - \lambda \frac{(2 - \sigma)^M - 1}{1 - \sigma} - c_3(k_1 k_2)^2 \left(\frac{11}{15} \right)^M.$$

Define the integer M by the inequality

$$c_3(k_1 k_2) \left(\frac{11}{15} \right)^M < \frac{1}{2} \leq c_3(k_1 k_2)^2 \left(\frac{11}{15} \right)^{M-1}.$$

Then

$$F(\sigma) \geq \frac{1}{2} - \frac{\lambda}{1 - \sigma} (2 - \sigma)^M.$$

Since $M < 8 \log k_1 k_2 + c_4$, then

$$(2 - \sigma)^M = e^{M \log(1 + 1 - \sigma)} < e^{M(1 - \sigma)} < c_5(k_1 k_2)^{8(1 - \sigma)}.$$

This completes the proof of the Lemma.   □

Proof of the Theorem. First, we shall prove the existence of an integer $k_0 = k_0(\varepsilon)$ such that if $k > k_0$ and $\sigma > 1 - 1/k^\varepsilon$ then

$$L(\sigma, \chi) \neq 0, \tag{10}$$

where χ is a real primitive character modulo k. This will immediately imply the Theorem.

We shall assume that there does not exist an integer k for which the function $L(s, \chi)$ has a zero in the interval $1 - \varepsilon/10 \leq \sigma < 1$. Denoting by $k_1 = k_1(\varepsilon)$ the smallest k that satisfies the condition $k^\varepsilon \geq 10/\varepsilon$, we obtain (10) for $k > k_1(\varepsilon)$.

We shall now assume existence of an integer k_1 and a real primitive character χ_1 modulo k_1 such that $L(s, \chi_1)$ has a zero $s = \sigma_1$ in the interval $[1 - \varepsilon/10, 1)$. Let k_2 be an arbitrary natural number greater than k_2, and let χ_2 be a real primitive character modulo k_2. Note that $\chi_2 \neq \chi_1$ since $k_2 > k_1$.

It follows from Lemma 2 and from the equation $L(\sigma_1, \chi_1) = 0$ that

$$0 = F(\sigma_1) > \frac{1}{2} - \frac{c\lambda}{1 - \sigma_1}(k_1 k_2)^{8(1 - \sigma_1)}, \quad 1 - \frac{\varepsilon}{10} \leq \sigma_1 < 1,$$

where

$$F(s) = \zeta(s)L(s, \chi_1)L(s, \chi_2)L(s, \chi_1\chi_2),$$

$$\lambda = L(1, \chi_1)L(1, \chi_2)L(1, \chi_1\chi_2).$$

Thus

$$L(1, \chi_1)L(1, \chi_2)L(1, \chi_1\chi_2) > c_1(1 - \sigma_1)(k_1 k_2)^{-0.8\varepsilon}.$$

Applying the estimates for $L(1, \chi_1)$ and $L(1, \chi_1\chi_2)$ from the Corollary to Lemma 9, Chapter VIII, we find that

$$L(1, \chi_2) \geq c_2(1 - \sigma_1)(k_1 k_2)^{-0.8\varepsilon}(\log k_1 k_2)^{-2}.$$

Now let $k_2 = k_2(\varepsilon, k_1, \sigma_1)$ be sufficiently large that

$$k_1^{-0.8\varepsilon}c_2(1 - \sigma_1)(\ln k_1 k_2)^{-2} > k_2^{-0.1\varepsilon}.$$

Then for all $k > k_2$ we shall have

$$L(1, \chi) > k^{-0.9\varepsilon},$$

where χ is a real primitive character modulo k. From this and from the estimates above for $L'(\sigma, \chi)$ we obtain

$$L(\sigma, \chi) - L(1, \chi) - (1 - \sigma)L'(\sigma_2, \chi) \geq k^{-0.9\varepsilon} - (1 - \sigma)c_3 \log^2 k > 0,$$

if $1 - 1/k^\varepsilon \leq \sigma < 1, k \geq k_3(\varepsilon) = k_3$. Consequently, we obtain (10) for $k > \max(k_1, k_3)$. This proves the Theorem. ☐

Remark. The constant $c = c(\varepsilon)$ in the Theorem is not effective, i.e., given $\varepsilon > 0$ it is not possible to find $c = c(\varepsilon)$. Therefore, every result that uses this theorem in an essential way is not effective (for example, Corollary 2 of Theorem 6).

§3. The Prime Number Theorem for Arithmetic Progressions

Applying the results of the preceding paragraph, we can obtain asymptotic formulas for $\psi(x; k, l)$ and $\pi(x; k, l)$.

Theorem 6. *For $x > 1$ we have*

$$\psi(x; k, l) = \frac{x}{\varphi(k)} - E_1 \frac{x^{\beta_1} \chi_1(l)}{\beta_1 \varphi(k)} + O(xe^{-c_0 \sqrt{\log x}}),$$

$$\pi(x; k, l) = \frac{\operatorname{Li} x}{\varphi(k)} - E_1 \frac{\chi_1(l)}{\varphi(k)} \int_2^x \frac{u^{\beta_1 - 1}}{\log u} \, du + O(xe^{-c_0' \sqrt{\log x}}),$$

where $E_1 = 1$ if there exists a real character χ_1 modulo k such that $L(s, \chi_1)$ has a real zero $\beta_1 > 1 - c/\log k$ and $E_1 = 0$ otherwise.

Proof. We shall assume that $k \le e^{\sqrt{\log x}}$. By Corollary 1 of Theorem 4, there exists at most one character χ_1 for which $E_1 = 1$. By formula (1)

$$\psi(x; k, l) = \frac{\psi(x)}{\varphi(k)} - \frac{E_1}{\varphi(k)} \chi_1(l) \psi(x, \chi_1) + \frac{1}{\varphi(k)} \sum_{\chi \ne \chi_0, \chi_1} \psi(x, \chi) \bar{\chi}(l) + O(\log^2 x).$$

Let $\chi \ne \chi_0, \chi_1$ and let χ^* be the primitive character modulo k_1 that induces χ, where $k_1 | k$. Then by Theorem 1 with $T = e^{\sqrt{\log x}}$

$$\psi(x, \chi^*) = \sum_{n \le x} \Lambda(n) \chi^*(n) = - \sum_{|\operatorname{Im} \rho| \le T} \frac{x^\rho}{\rho} + O(xe^{-0,5\sqrt{\log x}}),$$

where the ρ are the nontrivial zeros of $L(s, \chi^*)$. By Theorem 2,

$$\operatorname{Re} \rho = \beta \le 1 - \frac{c_1}{\log kT} \le 1 - \frac{c_2}{\sqrt{\log x}}.$$

Therefore,

$$|\psi(x, \chi^*)| \le \sum_{|\operatorname{Im} \rho| \le T} \frac{x^\beta}{\sqrt{\beta^2 + \gamma^2}} + c_3 x e^{-0.5\sqrt{\log x}} \le xe^{-c_0\sqrt{\log x}}.$$

Here we used Corollary 1 of Theorem 3, Chapter VIII.
Thus,

$$\psi(x, \chi) = \psi(x, \chi^*) + \theta_1 \log^2 x = O(xe^{-c_0\sqrt{\log x}}).$$

Now let $\chi = \chi_1$. Then

$$\psi(x, \chi_1) = - \sum_{\substack{|\operatorname{Im} \rho| \le T \\ \rho \ne \beta_1}} \frac{x^\rho}{\rho} + O(xe^{-0,5\sqrt{\log x}}),$$

since

$$\operatorname{Re} \rho = \beta \le 1 - \frac{c_1}{\log kT} \le 1 - \frac{c_2}{\sqrt{\log x}}.$$

Estimating the last sum for $\rho \ne \beta_1$ just as was done above, we find that

$$\psi(x; k, l) = \frac{\psi(x)}{\varphi(k)} - E_1 \frac{\chi_1(l)}{\beta_1} \frac{x^{\beta_1}}{\varphi(k)} + O(xe^{-c_0\sqrt{\log x}}).$$

Since

$$\psi(x) = x + O(xe^{-c_0\sqrt{\log x}}),$$

we obtain the first part of the Theorem.

We obtain the second part of the Theorem from the first by partial summation (Lemma C of Chapter I):

$$\pi(x; k, l) = \sum_{2 < n \le x} \frac{\Lambda(n)\alpha(n)}{\log n} + O(\sqrt{x}\log^2 x),$$

where

$$\alpha(n) = \alpha(n; k, l) = \begin{cases} 1, & \text{if } n \equiv l \pmod{k}; \\ 0, & \text{if } n \not\equiv l \pmod{k}. \end{cases}$$

Hence,

$$\pi(x; k, l) = \int_2^x \frac{\psi(u; k, l)}{u \log^2 u} du + \frac{\psi(x; k, l)}{\log x} + O(\sqrt{x}\log^2 x)$$

$$= \frac{1}{\varphi(k)} \int_2^x \frac{du}{\log^2 u} - E_1 \frac{\chi_1(l)}{\beta_1 \varphi(k)} \int_2^x \frac{u^{\beta_1 - 1} du}{\log^2 u}$$

$$+ \frac{x}{\varphi(k)\log x} - E_1 \frac{\chi_1(l) x^{\beta_1}}{\beta_1 \varphi(k)\log x} + O(xe^{-c_0'\sqrt{\log x}})$$

$$= \frac{\operatorname{Li} x}{\varphi(k)} - E_1 \frac{\chi_1(l)}{\varphi(k)} \int_2^x \frac{u^{\beta_1 - 1}}{\log u} du + O(xe^{-c_0'\sqrt{\log x}}),$$

which is exactly what we had to prove.   □

From this last result, from Theorems 3 and 5, and from Corollaries 2 and 3 of Theorem 4, we can deduce three corollaries about the distribution of primes in arithmetic progressions.

Corollary 1. *Let* $1 \le k \le (\log x)^{2-\varepsilon}$, *where* $0 < \varepsilon < 1/2$. *Then*

$$\psi(x; k, l) = \frac{x}{\varphi(k)} + O(xe^{-c(\log x)^{\varepsilon/3}}),$$

$$\pi(x; k, l) = \frac{\operatorname{Li} x}{\varphi(k)} + O(xe^{-c'(\log x)^{\varepsilon/3}}).$$

Proof. By Theorem 3,

$$\beta_1 \le 1 - \frac{c}{\sqrt{k}\log^4 k}.$$

Therefore,

$$x^{\beta_1} \leq x e^{-(c\log x/\sqrt{k}\log^4 k)} = O(xe^{-c(\log \alpha)^{t/3}}),$$

which is what we had to prove.   □

Corollary 2. *For any fixed $A > 1$ and $1 \leq k \leq (\ln x)^A$ we have the following asymptotic formulae:*

$$\psi(x; k, l) = \frac{x}{\varphi(k)} + O(xe^{-c_1\sqrt{\log x}}),$$

$$\pi(x; k, l) = \frac{\text{Li} x}{\varphi(k)} + O(xe^{-c_1\sqrt{\log x}}),$$

where $c_1 = c_1(A) > 0$.

Proof. By Lemma 5 for any $\varepsilon > 0$

$$\beta_1 \leq 1 - c(\varepsilon)/k^\varepsilon.$$

Let $\varepsilon = 1/2A$. Then

$$x^{\beta_1} \leq x e^{-\frac{c(\varepsilon)\log x}{k^\varepsilon}} \leq x e^{-c(\varepsilon)(\log x)^{1-A\varepsilon}} = x e^{-c_1\sqrt{\log x}},$$

which is that we had to prove.   □

Remark. The constant $c_1 = c_1(A)$ is not effective, i.e. $c_1 = c_1(A)$ cannot be computed for given A (cf. Theorem 5).

Corollary 3. *Let $x \geq y > 3$, and consider all k not exceeding y. Then, except, perhaps, for certain "special" moduli k, which are multiples of some k_0 such that $k_0 \geq c \log^2 y (\log \log y)^{-8}$, the following asymptotic formulae hold for all other k:*

$$\psi(x; k, l) = \frac{x}{\varphi(k)} + O(xe^{-c_2\sqrt{\log x}}) + O(xe^{-c_2(\log x/(\log y))}),$$

$$\pi(x; k, l) = \frac{\text{Li} x}{\varphi(k)} + O(xe^{-c_2'\sqrt{\log x}}) + O(xe^{-c_2'(\log x/\log y)}).$$

The proof follows from Corollaries 2 and 3 to Theorem 4.   □

Exercises

1. a) Let χ be a primitive character modulo k, Re $s = \sigma > 0$, $Z \geq k(|t| + 1)$. Then there is the following simple approximation:

$$L(s, \chi) = \sum_{n \leq Z} \frac{\chi(n)}{n^s} + O(kZ^{-\sigma}).$$

b) Let χ be a primitive character modulo k, $|\arg \eta| < \pi/2$. Then, with the notation of Chapter VIII, we have

$$\Gamma\left(\frac{s+\delta}{2}\right) L(s,\chi) = \sum_{n=1}^{\infty} \frac{\chi(n)}{n^s} \Gamma\left(\frac{s+\delta}{2}, \frac{\pi\eta n^2}{k}\right)$$

$$+ \frac{i^\delta \sqrt{k}}{\tau(\chi)} \left(\frac{k}{\pi}\right)^{0.5-s} \sum_{n=1}^{\infty} \frac{\bar{\chi}(n)}{n^{1-s}} \Gamma\left(\frac{1-s+\delta}{2}, \frac{\pi n^2}{k\eta}\right),$$

where $\Gamma(z,x)$ is the incomplete Gamma function (cf. Chapter IV, exercise 1).

2. Let χ be a primitive character modulo k, $k \le Q$. Then for any a_n we have the inequality

$$\sum_{k \le Q\chi} {\sum_{\bmod k}}' \left| \sum_{n=M+1}^{M+N} a_n \chi(n) \right|^2 \le c(Q^2+N) \sum_{n=M+1}^{M+N} |a_n|^2.$$

3. Let $2 \ge \operatorname{Re} s_\chi = \sigma_\chi \ge 0$, $\operatorname{Im} s_\chi = t_\chi$, $A \le t_\chi \le A+1$. Under the assumptions of exercise 2, we have (with $L = \ln N$)

$$\sum_{k \le Q\chi} {\sum_{\bmod k}}' \left| \sum_{n=M+1}^{M+N} a_n \chi(n) n^{-s}\chi \right|^2 \le c(Q^2+N)L^2 \sum_{n=M+1}^{M+N} |a_n|^2.$$

4. Let $\operatorname{Re} s > 0$ and let χ be a character modulo k. Prove that for $s = \rho$, $L(\rho, \chi) = 0$, $\operatorname{Re} \rho > 0$, $Z \ge k(|t|+1)$, $1 \le X < Y < Z$ one of the following inequalities holds:

$$1 \le c_2 \left| \sum_{X \le n \le X^2} a(n)\chi(n)n^{-\rho} \right|^4 = \kappa_1;$$

$$1 \le c_2 \left| \sum_{X^2 < n \le XY} a(n)\chi(n)n^{-\rho} \right|^2 = \kappa_2;$$

$$1 \le c_2 \left| \sum_{n \le X} \mu(n)\chi(n)n^{-\rho} \right|^{4/3} \left| \sum_{Y < n \le Z} \chi(n)n^{-\rho} \right|^{4/3} = \kappa_3;$$

$$1 \le c_2 k^2 Z^{-2\sigma} \left| \sum_{n \le X} \mu(n)\chi(n)n^{-\rho} \right|^2 = \kappa_4,$$

where $|a_n| \le \tau(n)$.

5. Let $N(\alpha, T, \chi)$ denote the number of zeros of $L(s,\chi)$ in the domain $\operatorname{Re} s \ge \alpha$, $|\operatorname{Im} s| \le T$. Then, for $0.5 \le \alpha \le 1$, $T \ge 2$, $Q \ge 1$ and under the conditions of exercise 2, we have

$$\sum_{k \le Q} {\sum_{\chi \bmod k}}' N(\alpha, T, \chi) \le cT(Q^2+QT)^{\frac{4(1-\alpha)}{3-2\alpha}} \log^{10}(Q+T).$$

6. a) For any $A > 0$ there exists a $B = B(A) > 0$ such that

$$\sum_{k \le \sqrt{X}(\ln X)^{-B}} \max_{(l,k)=1} \left| \psi(X;k,l) - \frac{X}{\varphi(k)} \right| \le c\frac{X}{(\ln X)^A},$$

where the constant $c = c(A) > 0$ cannot be effectively computed.

b) There exists a constant $B > 0$ such that

$$\sum_{k \le \sqrt{X}(\ln X)^{-B}} \max_{(l,k)=1} \left| \psi(X; k, l) - \frac{X}{\varphi(k)} \right| \le c_1 \frac{X}{(\ln X)^{2-\varepsilon}},$$

where $c_1 > 0$ is an effectively computable constant.

7. Let $k = p^n$, where $p \ge 3$ is a prime number, and let s and m be natural numbers. Suppose that $s \le n - 1$ and $n - s \le sm < n + s - 1$, and let $\operatorname{ind} v$ denote the index of the integer v modulo k. Then

$$\frac{\operatorname{ind}(1 + p^s u)}{p - 1} \equiv a_1 p^s u + \frac{1}{2} a_2 (p^s u)^2 + \ldots + \frac{1}{m} a_m (p^s u)^m \pmod{p^{n-1}},$$

where $(a_1, p) = (a_2, p) = \ldots = (a_m, p) = 1$ and the number $v^{-1} \pmod{p^{n-1}}$ is defined by the congruence $vv_1 \equiv 1 \pmod{p^{n-1}}$.

8. Let χ be an arbitrary non-principal character modulo $k = p^n$, where $p \ge 3$ is a fixed prime number. Then for $1 \le r \le 0.5n$, $N^r = k$, we have the estimate

$$\left| \sum_{m \le N} \chi(m) \right| \le c_1 N^{1 - c/r^2}.$$

9. Under the conditions of exercise 8, the function $L(s, \chi) \ne 0$ in the domain

$$|\operatorname{Im} s| < e^{c_2 (\ln \ln k)^2}, \quad \operatorname{Re} s = \sigma > 1 - \frac{c_3}{(\ln k)^{2/3} (\ln \ln k)^2}.$$

10. Prove that for $k = p^n$, where $p \ge 3$ is a fixed prime number and $k \le x^{1/9}$, $x \to +\infty$, we have the asymptotic formula

$$\psi(x; k, l) = \frac{x}{\varphi(k)} \{1 + O(e^{-c(\ln \ln x)^2})\}.$$

Chapter X. The Goldbach Conjecture

In this chapter we investigate the question of the representation of an odd integer N as the sum of three prime numbers (the Goldbach Conjecture). We shall prove I.M. Vinogradov's theorem on the asymptotic formula for the number of representations of N as the sum of three primes, from which it will follow that every sufficiently large odd number can be written as the sum of three primes.

We shall first give a simple, but non-effective proof (Theorem 3), and then an effective one (Theorem 4).

§1. Auxiliary Statements

We shall derive an analytic formula for the number of representations of an integer N as the sum of three prime numbers.

Lemma 1. *Let $J(N)$ denote the number of solutions in prime numbers p_1, p_2, p_3 of the equation $N = p_1 + p_2 + p_3$. Then*

$$J(N) = \int_0^1 S^3(\alpha) e^{-2\pi i \alpha N} \, d\alpha, \tag{1}$$

where

$$S(\alpha) = \sum_{p \leq N} e^{2\pi i \alpha p}.$$

Proof. If m is an integer different from zero, then

$$\int_0^1 e^{2\pi i \alpha m} \, d\alpha = \frac{e^{2\pi i \alpha m}}{2\pi i m}\bigg|_0^1 = 0.$$

Consequently,

$$\int_0^1 e^{2\pi i \alpha m} \, d\alpha = \begin{cases} 1 & \text{if } m = 0 \\ 0 & \text{if } m \text{ is an integer, } m \neq 0. \end{cases}$$

From this we obtain

$$J(N) = \sum_{p_1, p_2, p_3 \le N} \int_0^1 e^{2\pi i \alpha (p_1 + p_2 + p_3 - N)} d\alpha = \int_0^1 (S(\alpha))^3 e^{-2\pi i \alpha N} d\alpha,$$

This completes the proof.   □

The essence of the circle method of G.H. Hardy, J.E. Littlewood, and S. Ramanujan, as expressed by I.M. Vinogradov, in terms of trigonometric sums, consists in the fact that one can isolate from $J(N)$ a conjectural main term in an asymptotic formula for $J(N)$ as $N \to \infty$. To do this, we partition the interval of integration $[0, 1)$ in (1) by certain rational numbers (the Farey fractions). The sum of the integrals over the intervals that correspond to fractions with small denominators will give the conjectural main term. We shall need a lemma on the approximation of real numbers by rationals.

Lemma 2. *Let* $\tau \ge 1$, *and let* α *be a real number. Then there exist relatively prime integers* a *and* q *with* $1 \le q \le \tau$ *such that*

$$\left| \alpha - \frac{a}{q} \right| \le \frac{1}{q\tau}.$$

Proof. Without loss of generality, we can assume that $0 \le \alpha < 1$. We consider the numbers $\{\alpha m\}$ for $m = 0, 1, \ldots, [\tau]$. They lie in the interval $[0, 1)$, and, consequently, there are two different values of m, say, m_1 and m_2, such that

$$\{\alpha m_1\} - \{\alpha m_2\} = \theta/\tau, \quad |\theta| \le 1,$$

and so

$$\alpha(m_1 - m_2) - [\alpha m_1] + [\alpha m_2] = \theta/\tau,$$

where $1 \le |m_1 - m_2| \le [\tau] \le \tau$. This implies the Lemma.   □

§2. The Circle Method for Goldbach's Problem

We shall now extract the conjectural main term in the asymptotic formula for the value of $J(N)$. From now on we shall assume that $N \ge N_0$, where N_0 is a sufficiently large, fixed positive number. We first transform $J(N)$.

Let A and B be two positive numbers. We shall choose concrete values for A and B later. Let $L = \ln N, \tau = N \cdot L^{-B}, Q = L^A$, and $\kappa \tau = 1$. Since the integrand in (1) has period 1, we have

$$J(N) = \int_{-\kappa}^{1-\kappa} S^3(\alpha) e^{-2\pi i \alpha N} d\alpha. \tag{2}$$

By Lemma 2, each α in the interval $[-\kappa, 1 - \kappa]$ can be represented in the form

$$\alpha = \frac{a}{q} + z, \quad 1 \leq q \leq \tau, \quad (a, q) = 1, \quad |z| \leq \frac{1}{q\tau}. \tag{3}$$

It is easy to see that $0 \leq a \leq q - 1$ in this representation. Moreover, $a = 0$ only if $q = 1$. Denote by E_1 the set of those α for which $q \leq Q$ in the representation (3), and let E_2 denote the set of the remaining α. The set E_1 consists of non-intersecting intervals. Indeed, E_1 consists of intervals $E(a, q)$ of the form

$$\frac{a}{q} - \frac{1}{q\tau} \leq \alpha \leq \frac{a}{q} + \frac{1}{q\tau}, \quad 0 \leq a < q, \quad (a, q) = 1, \quad q = 1, 2, \ldots, |Q|.$$

If $E(a, q)$ and $E(a_1, q_1)$ are two different intervals of E_1, i.e. if $(a - a_1)^2 + (q - q_1)^2 \neq 0$, then the distance between the centers of these intervals is equal to

$$\left| \frac{a}{q} - \frac{a_1}{q_1} \right| \geq \frac{1}{qq_1},$$

and the sum of their radii is equal to

$$\frac{1}{q\tau} + \frac{1}{q_1\tau} < \frac{1}{qq_1}.$$

Consequently, $E(a, q)$ and $E(a_1, q_1)$ do not intersect.

Denote by J_1 the integral over the set E_1, and by J_2 the integral over set E_2, i.e.

$$J_1 = J_1(N) = \int_{E_1} S^3(\alpha) e^{-2\pi i \alpha N} d\alpha,$$

$$J_2 = J_2(N) = \int_{E_2} S^3(\alpha) e^{-2\pi i \alpha N} d\alpha.$$

Then $J = J_1 + J_2$. The object of this section is to obtain an asymptotic formula for the value of J_1. We shall need the following.

Lemma 3. *Let $\alpha \in E_1$ be of the form* (3). *Then*

$$S(\alpha) = \frac{\mu(q)}{\varphi(q)} M(z) + O(Ne^{-c\sqrt{L}}),$$

where

$$M(z) = \sum_{n=3}^{N} \frac{e^{2\pi i z n}}{\log n} = \int_3^N \frac{e^{2\pi i z u}}{\log u} du + O(1) + O(|z|N).$$

Proof. For any n in the interval $\sqrt{N} < n \leq N$, Corollary 2 of Theorem 6, Chapter IX gives

$$\pi(n; q, l) = \frac{\mathrm{Li}n}{\varphi(q)} = O(ne^{-c_1\sqrt{L}}).$$

Therefore,

$$S(\alpha) = S\left(\frac{a}{q} + z\right) = \sum_{\sqrt{N} < p \le N} e^{2\pi i \frac{ap}{q}} e^{2\pi i z p} + O(\sqrt{N})$$

$$= \sum_{\substack{l=1 \\ (l,q)=1}}^{q} e^{2\pi i \frac{al}{q}} T(l) + O(\sqrt{N}), \tag{4}$$

where

$$T(l) = \sum_{\substack{p \equiv l (\mathrm{mod}\, q) \\ \sqrt{N} < p \le N}} e^{2\pi i z p} = \sum_{\sqrt{N} < n \le N} (\pi(n; q, l) - \pi(n-1; q, l)) e^{2\pi i z n}.$$

We apply partial summation (Lemma C, Chapter I) to the last sum, with

$$c_n = \pi(n; q, l) - \pi(n-1; q, l), \quad f(u) = e^{2\pi i z u}.$$

Using the asymptotic formula for $C(u)$,

$$C(u) = \sum_{\sqrt{N} < n \le u} c_n = \frac{1}{\varphi(q)} \mathrm{Li}\, u + O(u e^{-c_1 \sqrt{L}}),$$

and the fact that, for $n \ge 3$,

$$\int_{n-1}^{n} \frac{e^{2\pi i z u}}{\log u}\, du = \frac{e^{2\pi i z n}}{\log n} + O(|z|) + O\left(\frac{1}{n \log^2 n}\right),$$

we obtain the Lemma from (1). $\qquad\square$

Remark. The constant in the symbol O is not effective, because Corollary 2 of Theorem 6, Chapter IX was used in the proof in an essential way.

Lemma 4. *The following formula holds for* J_1:

$$J_1 = \sigma\kappa + O(N^2 L^{-A-1}) + O(N^2 L^{-2B+2A}),$$

where

$$\sigma = \sum_{q=1}^{\infty} \gamma(q); \quad \gamma(q) = \frac{\mu(q)}{\varphi^3(q)} \sum_{\substack{a=1 \\ (a,q)=1}}^{q} e^{-2\pi i \frac{a}{q} N},$$

$$\kappa = \int_{-0.5}^{+0.5} M^3(z) e^{-2\pi i z N}\, dz; \quad M(z) = \sum_{n=3}^{N} \frac{e^{2\pi i z n}}{\log n}.$$

Proof. By definition,

$$J_1 = \int_{E_1} S^3(\alpha) e^{-2\pi i \alpha N}\, d\alpha = \sum_{q \le Q} \sum_{\substack{a=0 \\ (a,q)=1}}^{q-1} I(a, q),$$

where

$$I(a, q) = \int_{-1/q\tau}^{+1/q\tau} S^3\left(\frac{a}{q} + z\right) e^{-2\pi i \left(\frac{a}{q} + z\right) N}\, dz.$$

By Lemma 3,

$$S\left(\frac{a}{q} + z\right) = \frac{\mu(q)}{\varphi(q)} M(z) + O(Ne^{-c\sqrt{L}})$$

and so

$$S^3\left(\frac{a}{q} + z\right) = \frac{\mu(q)}{\varphi^3(q)} M^3(z) + O(N^3 e^{-c\sqrt{L}}).$$

Thus, for $I(a, q)$ we find that

$$I(a, q) = \frac{\mu(q)}{\varphi^3(q)} e^{-2\pi i \frac{a}{q} N} \int_{-1/q\tau}^{+1/q\tau} M^3(z) e^{-2\pi i z N} dz + O(N^2 L^B q^{-1} e^{-c\sqrt{L}}).$$

The integral in the last formula is nearly the same as the integral κ. We have

$$\int_{-1/q\tau}^{+1/q\tau} M^3(z) e^{-2\pi i z N} dz = \int_{-0.5}^{+0.5} M^3(z) e^{-2\pi i z N} dz + R = \kappa + R,$$

where

$$|R| \le 2 \int_{+1/q\tau}^{+0.5} |M(z)|^3 dz.$$

We estimate $|M(z)|$ for $0 < |z| \le 1/2$. Applying Lemma C of Chapter I (partial summation), we find that

$$|M(z)| = \left| -\int_2^N \mathbb{C}(u) d\frac{1}{\log u} + \mathbb{C}(N)\frac{1}{\log N} \right|, \quad |\mathbb{C}(u)| = \left| \sum_{2 < n \le u} e^{2\pi i z n} \right| \ll \frac{1}{|z|};$$

$$|M(z)| = O(1/|z|).$$

Therefore,

$$|R| \ll \int_{+1/q\tau}^{+0.5} \frac{dz}{z^3} \ll q^2 \tau^2 \ll N^2 L^{-2B+2A}.$$

Thus, we find that

$$I(a, q) = \frac{\mu(q)}{\varphi^3(q)} e^{-2\pi i \frac{a}{q} N} \kappa + O\left(\frac{1}{\varphi^3(q)} N^2 L^{-2B+2A}\right);$$

$$J_1 = \kappa \sum_{q \le Q} \frac{\mu(q)}{\varphi^3(q)} \sum_{\substack{a=0 \\ (a,q)=1}}^{q-1} e^{-2\pi i \frac{a}{q} N} + O(N^2 L^{-2B+2A}).$$

We transform the double sum in the last equation just as we earlier transformed the integral with respect to z. We have

$$\sum_{q \le Q} \frac{\mu(q)}{\varphi^3(q)} \sum_{\substack{a=0 \\ (a,q)=1}}^{q-1} e^{-2\pi i \frac{a}{q} N} = \sum_{q=1}^{\infty} \frac{\mu(q)}{\varphi^3(q)} \sum_{\substack{a=0 \\ (a,q)=1}}^{q-1} e^{-2\pi i \frac{a}{q} N} - R_1 = \sigma - R_1,$$

where

$$|R_1| < \left| \sum_{\substack{q > Q}} \frac{\mu(q)}{\varphi^3(q)} \sum_{\substack{a=0 \\ (a,q)=1}}^{q-1} e^{-2\pi i \frac{a}{q} N} \right| \leq \sum_{q > Q} \frac{1}{\varphi^2(q)} \ll \int_Q^\infty \frac{(\log\log u)^2}{u^2} \, du \ll L^{-A+1}$$

Consequently,

$$J_1 = \sigma\kappa + O(\kappa L^{-A+1}) + O(N^2 L^{-2B+2A}).$$

Finally,

$$|\kappa| = \left| \int_{-0.5}^{+0.5} M^3(z) e^{-2\pi i z N} \, dz \right| \leq N \int_{-0,5}^{+0,5} |M(z)|^2 \, dz$$

$$= N \sum_{3 \leq n \leq N} \frac{1}{\log^2 n} \ll N^2 L^{-2}.$$

Thus, we get the formula

$$J_1 = \sigma\kappa + O(N^2 L^{-A-1}) + O(N^2 L^{-2B+2A}),$$

which is what we had to prove.   □

We shall now study the values of κ and σ more closely.

Lemma 5. *We have the identity*

$$\kappa = \kappa(N) = \frac{N^2}{2\log^3 N} + O\left(\frac{N^2}{\log^4 N}\right).$$

Proof. Let

$$M_0(z) = \sum_{n=3}^{N} \frac{e^{2\pi i z n}}{\log N}.$$

Then

$$|M(z) - M_0(z)| \leq \sum_{n=3}^{N} \left(\frac{1}{\log n} - \frac{1}{\log N}\right)$$

$$\leq \int_{2}^{N} \left(\frac{1}{\log u} - \frac{1}{\log N}\right) du = O\left(\frac{N}{\log^2 N}\right).$$

Further, setting

$$\kappa_0 = \kappa_0(N) = \int_{-0.5}^{+0.5} M_0^3(z) e^{-2\pi i z N} \, dz,$$

we find that

$$|\kappa - \kappa_0| \ll \frac{N}{\log^2 N} \int_{-0.5}^{+0.5} (|M(z)|^2 + |M_0(z)|^2) \, dz \ll \frac{N^2}{\log^4 N}.$$

Consequently,

$$\kappa = \kappa_0 + O\left(\frac{N^2}{\log^4 N}\right) = \frac{1}{\log^3 N} I_0(N) + O\left(\frac{N^2}{\log^4 N}\right),$$

where $I_0(N)$ is the number of solutions of the equation

$$n_1 + n_2 + n_3 = N, \quad 3 \leq n_1, n_2, n_3 \leq N - 6.$$

For fixed n_3, with $3 \leq n_3 \leq N - 6$, the equation

$$n_1 + n_2 = N - n_3, \quad 3 \leq n_1, \quad n_2 \leq N - 6$$

has $N - n_3 - 5$ solutions. Therefore,

$$I_0(N) = \sum_{n_3 = 3}^{N-6} (N - n_3 - 5) = \frac{N^2}{2} + O(N).$$

Thus,

$$\kappa(N) = \frac{N^2}{2 \log^3 N} + O\left(\frac{N^2}{\log^4 N}\right),$$

which is what we had to prove.   □

Lemma 6. *We have the identity*

$$\sigma = \sigma(N) = \prod_{p \backslash N} \left(1 - \frac{1}{(p-1)^2}\right) \prod_{p \times N} \left(1 + \frac{1}{(p-1)^3}\right).$$

Proof. We shall first show that the sum $T(q)$

$$T(q) = \sum_{\substack{a = 1 \\ (a, q) = 1}}^{q} e^{-2\pi i \frac{a}{q} N},$$

is a multiplicative function of q. Let $q = q_1 q_2$, where $(q_1, q_2) = 1$. Then

$$T(q_1 q_2) = \sum_{\substack{a = 1 \\ (a, q) = 1}}^{q_1 q_2} e^{-2\pi i \frac{a}{q} N} = \sum_{\substack{a_1 = 1 \\ (a_1, q_1) = 1}}^{q_1} \sum_{\substack{a_2 = 1 \\ (a_2, q_2) = 1}}^{q_2} e^{-2\pi i \frac{a_2 q_1 + a_1 q_2}{q_1 q_2} N}$$

$$= T(q_1) T(q_2).$$

From this follows the multiplicativity of $T(q)$ and $\gamma(q)$. Furthermore, since

$$|\gamma(q)| \leq 1/\varphi^2(q),$$

then

$$\prod_{p \leq X} (1 + \gamma(p) + \gamma(p^2) + \ldots) = \sum_{q \leq X} \gamma(q) + O\left(\sum_{q > X} \frac{1}{\varphi^2(q)}\right)$$

$$= \sum_{q \leq X} \gamma(q) + O\left(\frac{\log \log X}{X}\right).$$

Taking the limit as $X \to +\infty$, we find that

$$\sigma = \prod_p (1 + \gamma(p) + \gamma(p^2) + \ldots).$$

From the definition of $\gamma(q)$ we obtain

$$\gamma(p) = \begin{cases} -\dfrac{1}{(p-1)^2}, & \text{if } p \mid N; \\[3mm] -\dfrac{1}{(p-1)^3}, & \text{if } p \times N; \end{cases}$$

$$\gamma(p^r) = 0, \quad \text{if } r \geq 2.$$

Thus,

$$\sigma = \prod_{p \backslash N}\left(1 - \frac{1}{(p-1)^2}\right) \prod_{p \times N}\left(1 + \frac{1}{(p-1)^3}\right)$$

$$= \prod_p \left(1 + \frac{1}{(p-1)^3}\right) \prod_{p \backslash N}\left(1 - \frac{1}{p^2 - 3p + 3}\right),$$

This completes the proof. $\qquad\qquad\qquad\qquad\qquad\qquad\qquad\qquad\square$

From this Lemma follows the fundamental result of this section.

Theorem 1. *The following asymptotic formula holds for J_1:*

$$J_1 = J_1(N) = \frac{N^2}{2(\log N)^3}\sigma + O\left(\frac{N^2}{(\log N)^4}\right),$$

where

$$\sigma = \sigma(N) = \prod_{p \backslash N}\left(1 - \frac{1}{(p-1)^2}\right) \prod_{p \times N}\left(1 + \frac{1}{(p-1)^3}\right).$$

Remarks.

1. The constant in the symbol O in the Theorem is not effective, since the proof makes essential use of Corollary 2 of Theorem 6, Chapter IX.

2. We shall obtain later (cf. §3) an asymptotic formula for J_1 with an effective constant in the symbol O.

3. For an odd integer N, the following obvious inequalities

$$\prod_{p \backslash N}\left(1 - \frac{1}{(p-1)^2}\right) > \prod_p\left(1 - \frac{1}{p^2}\right) = \frac{6}{\pi^2}, \quad \prod_{p \times N}\left(1 + \frac{1}{(p-1)^3}\right) > 2$$

imply that

$$\sigma(N) > 1.$$

In order to obtain an asymptotic formula for $J = J(N)$, we need to estimate J_2. For this we shall need an estimate for $|S(\alpha)|$ for α belonging to the set E_2.

§3. Linear Trigonometric Sums with Prime Numbers

We shall prove Vinogradov's estimate for linear trigonometric sums over the prime numbers. As a Corollary of this Theorem and Theorem 1, we shall obtain the asymptotic formula for the number of representations of an odd integer N as the sum of three prime numbers.

Theorem 2. *Let*

$$H = e^{0.5\sqrt{\log N}}, \quad \alpha = \frac{a}{q} + \frac{\theta}{q^2}, \quad (a, q) = 1, \quad |\theta| \leq 1, \quad 1 < q \leq N;$$

$$S = S(\alpha) = \sum_{p \leq N} e^{2\pi i \alpha p}.$$

Then

$$S \ll N (\log N)^3 \Delta,$$

where

$$\Delta = \frac{1}{H} + \sqrt{\frac{1}{q} + \frac{q}{N}}.$$

Proof. We set

$$P = \prod_{p \leq \sqrt{N}} p.$$

Using the properties of the Möbius function, we find

$$\sum_{\substack{n=1 \\ (n,P)=1}}^{N} e^{2\pi i \alpha n} = \sum_{d \backslash P} \mu(d) S(d),$$

$$S(d) = \sum_{0 < m \leq Nd^{-1}} e^{2\pi i \alpha m d}.$$

Therefore,

$$S = S_0 - S_1 + O(\sqrt{N}), \tag{5}$$

where

$$S_0 = \sum\sum_{d_0 m \leq N} e^{2\pi i \alpha m d_0}, \quad \mu(d_0) = +1,$$

$$S_1 = \sum\sum_{m d_1 < N} e^{2\pi i \alpha m d_1}, \quad \mu(d_1) = -1.$$

The sums S_0 and S_1 can be estimated in the same way. We estimate S_0. We partition the interval $0 < m \leq N$ into $\ll \log N$ intervals of the form

$M < m \le M'$, $M' \le 2M$, and we consider the sum

$$S(M) = \sum_{\substack{md_0 \le N \\ M < m \le M'}} e^{2\pi i\alpha m d_0}. \tag{6}$$

If $M \ge H$, then, applying Lemma 4 of Chapter VI, we find

$$S(M) = \sum_{d_0 \le NM^{-1}} \sum_{M < m \le \min\left(M', \frac{N}{d_0}\right)} e^{2\pi i\alpha m d_0}$$

$$\ll \sum_{d_0 \le NM^{-1}} \min\left(\frac{N}{d_0}, \frac{1}{\|\alpha d_0\|}\right) \le \sum_{n \le NM^{-1}} \min\left(\frac{N}{n}, \frac{1}{\|\alpha n\|}\right)$$

$$\le \sum_{0 < n \le 0,5q} + \sum_{0,5q < n \le 1,5q} + \ldots + \sum_{(r-0,5)q < n \le (r+0,5)q}, \tag{7}$$

where $r \le NM^{-1}q^{-1}$. Let k be the least nonnegative residue of the number an modulo q for $1 \le n < 0.5q$. Then

$$\|\alpha n\| = \left\|\frac{an}{q} + \frac{0n}{q^2}\right\| = \left\|\frac{k + 0,5\theta_1}{q}\right\|, \quad |\theta_1| \le 1.$$

For this, setting

$$u = \begin{cases} k, & \text{if } k \le 0,5q; \\ q - k, & \text{if } k > 0,5q, \end{cases}$$

we find that

$$\|\alpha n\| \ge \frac{u - 0.5}{q}.$$

Therefore, the first summand in (7) is

$$\le q \sum_{0 < u \le 0,5q} \frac{1}{u - 0.5} \ll q \log q.$$

We apply Lemma 5 of Chapter VI to the remaining summands in (7), and obtain

$$S(M) \ll q \log q + \sum_{l=1}^{r} \left(\frac{N}{(l - 0.5)q} + q \log q\right)$$

$$\ll q \log q + Nq^{-1} \log N + NM^{-1} \log q \ll N(\log N)\left(\frac{q}{N} + \frac{1}{q} + \frac{1}{H}\right). \tag{8}$$

Now let $M < H$. We represent the sum $S(M)$ in the form

$$S(M) = \sum_{M < m \le M'} \sum_{d_0 \le Nm^{-1}} e^{2\pi i\alpha m d_0}.$$

We denote by the letter δ_k each d_0 that has exactly k prime factors greater than

H^2. If k_0 is the maximum value of k for $d_0 \le N$, then $2^{k_0} \le N$, i.e. $k_0 \ll \log N$. We have

$$S(M) = \sum_{k=0}^{k_0} S_k(M),$$

$$S_k(M) = \sum_{M < m \le M'} \sum_{\delta_k \le Nm^{-1}} e^{2\pi i \alpha m \delta_k}.$$

We estimate $S_0(M)$. Let κ be the number of prime factors of δ_0, where $\delta_0 > NM^{-1}H^{-1}$. Then

$$H^{2\kappa} > NH^{-2}; \quad (2\kappa + 2)0.5\sqrt{\log N} > \log N;$$

$$\chi > \sqrt{\log N} - 1, \quad \tau(\delta_0) > 2^{\sqrt{\log N} - 1}.$$

Applying the trivial inequality

$$\sum_{n \le x} \tau(n) = \sum_{n \le x} \left[\frac{x}{n}\right] \ll x \log x,$$

we have

$$S_0(M) \ll \sum_{M < m \le M'} \left(\sum_{\delta_0 \le NM^{-1}H^{-1}} 1 + \sum_{NM^{-1}H^{-1} < \delta_0 \le Nm^{-1}} \frac{\tau(\delta_0)}{2^{\sqrt{\log N}}} \right)$$

$$\ll M \left(\frac{N}{MH} + \frac{N \log N}{M \cdot 2^{\sqrt{\log N}}} \right) \ll \frac{N}{H}.$$

We estimate $S_k(m)$, $k > 0$. Compare $S_k(m)$ with the sum

$$T_k = \sum_{M < m \le M'} \sum_{pt \le Nm^{-1}} e^{2\pi i \alpha m p t},$$

where p runs through the prime numbers in the interval $H^2 < p \le \sqrt{N}$, and t runs through the values d_1 having exactly $k - 1$ prime factors exceeding H^2. Let $k > 1$. For the terms in the sum T_k with $(p, t) = p$ we have

$$\ll \sum_{M < m \le M'} \sum_{H^2 < p \le \sqrt{N}} \frac{NM^{-1}}{p^2} \ll \frac{N}{H}.$$

The remaining terms of the sum T_k are just the terms in the sum $S_k(M)$. Moreover, each term of the sum $S_k(m)$ occurs in T_k exactly k times. Therefore,

$$S_k(M) = \frac{1}{k} T_k + 0\left(\frac{N}{kH}\right).$$

The last equation also holds for $k = 1$. We estimate T_k. Denote $mp = u$. We partition the interval

$$MH^2 < u \le M'\sqrt{N}$$

into $\ll \log N$ intervals $U < u \le U'$, $U < U' \le 2U$, and let

$$T_k(U) = \sum_{U < u \le U'}' \sum_{ut \le N} e^{2\pi i a u t}.$$

Applying Lemma 5 of Chapter VI, we obtain

$$|T_k(U)|^2 \le U \sum_{u=U+1}^{2U} \left| \sum_{ut \le N} e^{2\pi i a u t} \right|^2$$

$$= U \sum_{t_1 \le NU^{-1}} \sum_{t_2 \le NU^{-1}} \sum_{U \le u \le \min(2U,\, N/t_1,\, N/t_2)} e^{2\pi i a u (t_1 - t_2)}$$

$$\ll U \sum_{t_1 \le NU^{-1}} \sum_{t_2 \le NU^{-1}} \min\left(U, \frac{1}{\|\alpha(t_1 - t_2)\|} \right)$$

$$\ll U \frac{N}{U} \left(\frac{N}{Uq} + 1 \right) (U + q \log q) \ll N^2 \left(\frac{1}{q} + \frac{U}{N} + \frac{1}{U} + \frac{q}{N} \right)$$

$$\times \log N \ll N^2 \left(\frac{1}{q} + \frac{q}{N} + \frac{1}{H^2} \right) \log N;$$

$$|T_k(U)| \ll N\sqrt{\log N} \left(\frac{1}{H} + \sqrt{\frac{1}{q} + \frac{q}{N}} \right);$$

$$|T_k| \ll N(\log N)^{3/2} \left(\frac{1}{H} + \sqrt{\frac{1}{q} + \frac{q}{N}} \right).$$

From this and from (8) we obtain

$$S(M) \ll |S_0(M)| + \sum_{k=1}^{k_0} \left(\frac{1}{k} |T_k| + \frac{N}{kH} \right)$$

$$\ll N(\log N)^{3/2} (\log\log N) \left(\frac{1}{H} + \sqrt{\frac{1}{q} + \frac{q}{N}} \right);$$

$$S \ll N(\log N)^3 \left(\frac{1}{H} + \sqrt{\frac{1}{q} + \frac{q}{N}} \right),$$

This completes the proof. $\qquad\square$

Theorem 3. *The following asymptotic formula holds for the number $J(N)$ of representations of an odd integer as the sum of three primes:*

$$J(N) = \sigma(N) \frac{N^2}{2(\log N)^3} + O\left(\frac{N^2}{(\log N)^4} \right),$$

$$\sigma(N) = \prod_p \left(1 + \frac{1}{(p-1)^3} \right) \prod_{p \backslash N} \left(1 - \frac{1}{p^2 - 3p + 3} \right) > 1. \qquad (9)$$

Proof. From Lemma 1, the formulas in §2, and Theorem 1 for $A = 15$, we have

$$J(N) = J_1(N) + J_2(N) = \sigma(N)\frac{N^2}{2(\log N)^3} + J_2(N) + O\left(\frac{N^2}{(\log N)^4}\right),$$

where

$$J_2(N) = \int_{E_2} S^3(\alpha) e^{-2\pi i \alpha N} d\alpha.$$

It follows from the definition of the set E_2 that for $\alpha \in E_2$ we have

$$\alpha = \frac{a}{q} + \frac{\theta}{q^2}, \quad (a, q) = 1,$$

$$|\theta| \le 1, \quad (\log N)^{15} < q < N(\log N)^{-20}.$$

By Theorem 2,

$$S(\alpha) \ll N(\log N)^{-4}, \quad \alpha \in E_2.$$

Therefore,

$$J_2(N) \ll \max_{\alpha \in E_2}|S(\alpha)|\int_0^1|S(\alpha)|^2 d\alpha \ll N^2(\log N)^{-5}.$$

This completes the proof.   □

Corollary (the Goldbach conjecture). *There exists an N_0 such that every odd integer $N > N_0$ is the sum of three prime numbers.*

According to the remarks after Theorem 1, the constant in the symbol O in formula (9) is not effective, and so the constant N_0 is not effective. In the next section we shall obtain an effective asymptotic formula for $J(N)$, and so the constant N_0 in the Corollary will also be effective.

§4. An Effective Theorem

We shall first obtain a nontrivial estimate for $S(\alpha)$, the trigonometric sum over the prime numbers, in the case when α has a rational approximation with small denominator.

Lemma 7. *Let $\varepsilon_0 > 0$ be a sufficiently small constant,*

$$\tau \ge Ne^{-\varepsilon_0\sqrt{\log N}}, \quad N_1 \ge Ne^{-\varepsilon_0\sqrt{\log N}},$$

$$\alpha = \frac{a}{q} + z; \quad (a, q) = 1, \quad 0 < q \le e^{\varepsilon_0\sqrt{\log N}}, \quad |z| \le \frac{1}{q\tau}.$$

Then

$$S(\alpha) = \sum_{N-N_1 < p \leq N} e^{2\pi i a p} \ll \frac{N_1 \log\log q}{\sqrt{q} \log N}.$$

Proof. By Theorem 6 of Chapter IX,

$$\pi(n;q,l) = \frac{\text{Li}\,n}{\varphi(q)} - E_1 \frac{\chi_1(l)}{\varphi(q)} \int\limits_2^n \frac{u^{\beta_1 - 1}}{\log u}\,du + O(ne^{-c'\sqrt{\log N}}), \quad \sqrt{N} \leq n \leq N.$$

Therefore, repeating the first part of the proof of Theorem 1 with the transformation of $S(a/q + z)$, we have

$$S(\alpha) = S\left(\frac{a}{q} + z\right) = \sum_{\substack{l=1 \\ (l,q)=1}}^{q} T(l)e^{2\pi i \frac{a}{q} l} + O(\sqrt{N}),$$

$$T(l) = \sum_{N-N_1 \leq n \leq N} (t(n) - t(n-1))e^{2\pi i z n}$$

$$+ O(Nc^{-c_1\sqrt{\log N}}) + O(N^2 e^{-c_1\sqrt{\log N}}|z|),$$

where

$$t(n) = \frac{\text{Li}\,n}{\varphi(q)} - E_1 \frac{\chi_1(l)}{\varphi(q)} \int\limits_2^n \frac{u^{\beta_1 - 1}}{\log u}\,du.$$

Thus,

$$S(\alpha) = \frac{\mu(q)}{\varphi(q)} \sum_{N-N_1 \leq n \leq N} \left(\int\limits_{n-1}^{n} \frac{du}{\log u}\right) e^{2\pi i z n}$$

$$- \frac{E_1}{\varphi(q)} \left(\sum_{l=1}^{q} \chi_1(l)e^{2\pi i \frac{a}{q} l}\right) \sum_{N-N_1 \leq n \leq N} \left(\int\limits_{n-1}^{n} \frac{u^{\beta_1 - 1}}{\log u}\,du\right) e^{2\pi i z n}$$

$$+ O(qNe^{-c_1\sqrt{\log N}}) + O(qNe^{-c_1\sqrt{\log N}}|z|). \tag{10}$$

Since χ_1 is a real character modulo q, then (cf. Ch. VIII, §1)

$$\left|\sum_{l=1}^{q} \chi_1(l)e^{2\pi i \frac{a}{q} l}\right|^2 = \frac{1}{\varphi(q)} \sum_{\substack{m=1 \\ (m,q)=1}}^{q} \left|\sum_{l=1}^{q} \chi_1(l)e^{2\pi i \frac{m}{q} l}\right|^2$$

$$\leq \frac{1}{\varphi(q)} \sum_{m=1}^{q} \sum_{\substack{l=1 \\ (l,q)=1}}^{q} \sum_{\substack{n=1 \\ (n,q)=1}}^{q} \chi_1(l)\chi_1(n)e^{2\pi i \frac{m}{q}(l-n)} \leq q.$$

Passing to the inequality in (10), we find

$$S(\alpha) \ll \frac{\text{Li}\,N - \text{Li}(N-N_1)}{\varphi(q)} + \frac{\sqrt{q}(\text{Li}\,N - \text{Li}(N-N_1))}{\varphi(q)}$$

$$+ qNe^{-c_1\sqrt{\log N}} + qN^2 c^{-e_1\sqrt{\log N}}|z| \ll \frac{N_1}{\log N} \cdot \frac{\log\log q}{\sqrt{q}},$$

which is what we had to prove.   □

Theorem 4. *We have the following asymptotic formula for the number $J(N)$ of representations of an odd integer N as the sum of three prime numbers:*

$$J(N) = \sigma(N)\frac{N^2}{2(\log N)^3} + O\left(\frac{N^2}{(\log N)^{3.4}}\right),$$

where

$$\sigma(N) = \prod_{p}\left(1 + \frac{1}{(p-1)^3}\right)\prod_{p\backslash N}\left(1 - \frac{1}{p^2 - 3p + 3}\right)$$

and the constant in the symbol O is effective.

Proof. We choose $\tau = N(\log N)^{-20}$. By Lemma 2, for

$$\alpha \in \left[-\frac{1}{\tau}, 1 - \frac{1}{\tau}\right], \quad \alpha = \frac{a}{q} + z, \quad 1 \le q \le \tau, \quad (a, q) = 1, \quad |z| \le \frac{1}{q\tau}. \quad (11)$$

Denote by E_1 the set of those α for which $q \le (\log N)^3$, and by E_2 the set of all other α. As before,

$$J = J_1 + J_2$$

where

$$J_1 = \int_{E_1} S^3(\alpha)e^{-2\pi i\alpha N}\,d\alpha, \quad J_2 = \int_{E_2} S^3(\alpha)e^{-2\pi i\alpha N}\,d\alpha.$$

We shall estimate J_2. If

$$q \ge (\log N)^{20}$$

in the representation (11), then by Theorem 2

$$S(\alpha) \ll N(\log N)^{-7}.$$

If

$$(\log N)^3 < q \le (\log N)^{20},$$

then by Lemma 7

$$S(\alpha) \ll N(\log N)^{-2.5(\log\log N)}.$$

Therefore,

$$J_2 \ll \max_{\alpha \in E_2}|S(\alpha)| \int_0^1 |S(\alpha)|^2\,d\alpha \ll N^2(\log N)^{-3.5}(\log\log N).$$

Now we compute J_1. First we consider the set of all q not exceeding y, where

$$y = e^{\log N/(\log\log N)^2}.$$

Except, perhaps, for "special" moduli q, which are multiples of some q_0,

$$q_0 \ge c\log^2 y(\log\log y)^{-8} \ge c\log^2 N(\log\log N)^{-12},$$

it follows from Corollary 3 of Theorem 6 in Chapter IX, for $\sqrt{N} \leq x \leq N$ and for all remaining moduli, that we have the asymptotic formula

$$\pi(x; q, l) = \frac{\text{Li} x}{\varphi(q)} + O(xe^{-c_1 (\log \log x)^2}).$$

We represent the integral J_1 as the sum of two integrals:

$$J_1 = J_1' + J_1'',$$

where the integration in J_1' is carried out over those α for which the modulus $q \leq (\log N)^3$ in representation (11) is not one of the "special" moduli, and the integration in J_1'' is carried out over those α for which the modulus $q \leq (\log N)^3$ in representation (11) is one of the "special" moduli. Repeating the proof of Theorem 1 for the non-special moduli, we obtain

$$J_1' = \frac{N^2}{2(\log N)^3} \sum_{q \leq (\log N)^3}' \gamma(q) + O\left(\frac{N^2}{\log^4 N}\right), \tag{12}$$

where the summation in the last sum runs over the q that are not special moduli, and

$$\gamma(q) = \frac{\mu(q)}{\varphi^3(q)} \sum_{\substack{a=1 \\ (a,q)=1}}^{q} e^{-2\pi i \frac{a}{q} N}.$$

We estimate J_1''. Let $D = [(\log N)^{30}]$, $A = ND^{-1}$. Then

$$S\left(\frac{a}{q} + z\right) = \sum_{s=1}^{D} \sum_{(s-1)A < p \leq sA} e^{2\pi i \left(\frac{a}{q} + z\right)p}$$

$$= \sum_{s=1}^{D} \sum_{(s-1)A < p \leq sA} e^{2\pi i \frac{a}{q} p} \cdot e^{2\pi i z s A} + O(|z|AN)$$

$$= \sum_{s=1}^{D} e^{2\pi i z s A} \sum_{(s-1)A < p \leq sA} e^{2\pi i \frac{a}{q} p} + O(Nq^{-1}(\log N)^{-10}).$$

This implies that

$$S^2\left(\frac{a}{q} + z\right) e^{-2\pi i \left(\frac{\alpha}{\tau} + z\right)N} = \sum_{s_1, s_2, s_3 = 1}^{D} e^{2\pi i z A(s_1 + s_2 + s_3 - D)} W(s_1, s_2, s_3)$$

$$+ O\left(\left|S\left(\frac{a}{q} + z\right)\right|^2 Nq^{-1}(\log N)^{-10}\right) + O(N^3 q^{-3}(\log N)^{-30}),$$

where

$$W(s_1, s_2, s_3) = \sum_{(s_1-1)A < p_1 \leq s_1 A} \sum_{(s_2-1)A < p_2 \leq s_2 A} \sum_{(s_3-1)A < p_3 \leq s_3 A} e^{2\pi i \frac{a}{q}(p_1 + p_2 + p_3 - N)}.$$

Thus, we obtain for J_1'' the following estimate:

$$J_1'' \ll \sum_{\substack{q \le (\log N)^3}}'' \sum_{\substack{a=1 \\ (a,q)=1}}^{q} \left(\frac{1}{q^\tau} \sum_{\substack{s_1,s_2,s_3=1 \\ s_1+s_2+s_3=D}}^{D} |W(s_1,s_2,s_3)| \right.$$

$$\left. + \sum_{\substack{s_1,s_2,s_3=1 \\ s_1+s_2+s_3 \ne D}}^{D} \frac{1}{|s_1+s_2+s_3-D|A} |W(s_1,s_2,s_3)| \right) + N^2(\log N)^{-10}.$$

We use Lemma 7 to estimate $|W(s_1,s_2,s_3)|$, and find

$$|W(s_1,s_2,s_3)| \ll \left(\frac{A \log\log N}{\sqrt{q} \log N} \right)^3.$$

Moreover, the number of solutions of the equation

$$s_1 + s_2 + s_3 - D = \lambda$$

does not exceed D^2, $\lambda \ll D$.

Therefore,

$$J_1'' \ll \sum_{\substack{q \le (\log N)^3}}'' \left(\frac{1}{\tau} D^2 \frac{A^3 (\log\log N)^3}{q^{3/2} (\log N)^3} + \frac{q}{A} D^2 \frac{A^3 (\log\log N)^4}{q^{3/2} (\log N)^3} \right)$$

$$+ N^2(\log N)^{-10} \ll N^2 (\log N)^{-10} + \frac{N^2 (\log\log N)^4}{(\log N)^3} \sum_{\substack{q \le (\log N)^3}}'' \frac{1}{\sqrt{q}},$$

where the summation in the last sum runs over "special" q. Therefore,

$$\sum_{\substack{q \le (\log N)^3}}'' \frac{1}{\sqrt{q}} \ll \frac{1}{\sqrt{q_0}} \sum_{\substack{m \le (\log N)(\log\log N)^{12}}} \frac{1}{\sqrt{m}}$$

$$\ll \frac{1}{\sqrt{q_0}} \sqrt{\log N} (\log\log N)^6 \ll \frac{(\log\log N)^{12}}{\sqrt{\log N}}.$$

Finally, we obtain

$$J_1'' \ll \frac{N^2 (\log\log N)^{16}}{(\log N)^{3.5}}. \tag{13}$$

From the definition of $\gamma(q)$ and of the "special" moduli q, it follows that

$$\sum_{\substack{q \le (\log N)^3}}'' \gamma(q) \ll \sum_{\substack{q \le (\log N)^3}}'' \frac{1}{\varphi^2(q)} \ll \sum_{\substack{q \le (\log N)^3}}'' \frac{(\log\log q)^3}{q^2}$$

$$\ll q_0^{-2} (\log\log\log N)^2 \ll (\log N)^{-4} (\log\log N)^{25};$$

We find from (13) and from the last estimate that

$$J_1'' = \frac{N^2}{2(\log N)^3} \sum_{\substack{q \le (\log N)^3}}'' \gamma(q) + O\left(\frac{N^2 (\log\log N)^{16}}{(\log N)^{3.5}} \right).$$

Combining this expression for J_1'' with (12), we obtain an asymptotic formula for J_1, and, consequently, for J:

$$J_1 = \frac{N^2}{2(\log N)^3} \sum_{q \le (\log N)^3} \gamma(q) + O\left(\frac{N^2(\log\log N)^{16}}{(\log N)^{3,5}}\right)$$

$$= \sigma(N)\frac{N^2}{2(\log N)^2} + O\left(\frac{N^2}{(\log N)^{3,4}}\right);$$

$$J = \sigma(N)\frac{N^2}{2(\log N)^3} + O\left(\frac{N^2}{(\log N)^{3,4}}\right),$$

which is what we had to prove.   □

Exercises

1. For fixed natural numbers n, m, k, derive an asymptotic formula for the number of solutions of the equation

$$np_1 + mp_2 + kp_3 = N$$

for prime numbers p_1, p_2, p_3.

2. Let $K(X)$ denote the number of even integers not exceeding X and not representable as the sum of three prime numbers. Prove that

$$K(X) = O(X(\ln X)^{-D})$$

for any fixed $D > 0$.

3. Let P be a positive integer. Let z run through the integers $z_1, \ldots, z_n$. Let S' denote the sum of the values of the function $f(z) \ge 0$, where z runs through the integers that are relatively prime to P. Let S_d denote the sum of the values of the function $f(z)$, where z runs through the integers that are multiples of d. Then for any integer $m > 0$ we have

$$S' \le \sum_{\substack{d\backslash P \\ \Omega(d) \le m}} \mu(d)S_d.$$

4. a) Let $k \le x^{0.9}$, $\ln b = \ln x \cdot (1000\ln\ln x)^{-1}$; $0 \le l < k$, $(l, k) = 1$. Let T denote the number of integers of the form $kn + 1, n = 0, 1, \ldots$, not divisible by any prime $\le b$ and not exceeding x. Then we have

$$T \le \frac{cx\ln\ln x}{\varphi(k)\ln x}.$$

b) Let $0 < \alpha < 1, k \le x^\alpha, x \ge x_0 > 0$. Then

$$\pi(x; k, l) \le \frac{cx\ln\ln x}{\varphi(k)\ln x}.$$

5. Prove that

$$\sum_{p \leq x} \tau(p - 1) = c_0 x + O\left(\frac{x \ln \ln x)^3}{\ln x}\right),$$

where $c_0 > 0$ is an absolute constant.

6. a) Let p be a prime number, $(k, p) = 1$, and let q denote a prime number. Then there exists an absolute constant $\gamma > 0$ such that

$$\sigma = \left| \sum_{q < p^\gamma} \left(\frac{q + k}{p}\right) \right| \leq c p^{\gamma - \delta},$$

where $\delta = \delta(\gamma) > 0$.

b). Under the conditions of a), the number of quadratic residues (or non-residues) modulo p of the form $q + k$, $q \leq p^\gamma$, is equal to

$$\frac{1}{2} \pi(p^\gamma) + O(p^{\tau - \delta}).$$

7. Let p be a prime number, $(k, p) = 1$. There exists an absolute constant $\gamma > 0$ such that

a) $\left| \sum_{\substack{n \leq p^\gamma \\ \mu(n) \neq 0}} \left(\frac{\mu(n)n + k}{p}\right) \right| \leq c p^{\gamma - \delta}, \quad \delta = \delta(\gamma) > 0;$

b) The number of quadratic residues (or nonresidues) modulo p of the form $\mu(n)n + k$, $\mu(n) \neq 0$, $n \leq p^\gamma$, is equal to

$$\frac{6}{\pi^2} p^\gamma + O(p^{\gamma - \delta}), \quad \delta = \delta(\gamma) > 0.$$

Chapter XI. Waring's Problem

In this chapter we study the representation of a natural number N as the sum of a fixed number of fixed powers of natural numbers, i.e. the question of the solvability in natural numbers $x_1, x_2, \ldots, x_k$ of the equation

$$x_1^n + x_2^n + \ldots + x_k^n = N, \tag{1}$$

where $n \geq 3$ and $k = k(n)$ (Waring's problem). Waring's problem generalizes Lagrange's theorem that every natural number is the sum of four squares.

We shall prove here two results of I. M. Vinogradov concerning $J_{k,n}(N)$ – the number of solutions of equation (1). The first concerns the asymptotic formula for $J_{k,n}(N)$ as $N \to \infty$, which will be obtained when the number k of summands is of order $n^2 \log n$. In particular, it follows from this that there exists an integer $k = k(n)$ for which (1) is solvable in nonnegative integers for all $N \geq 1$. The second result is an upper bound for the smallest k as a function of n for which equation (1) is solvable for all sufficiently large N. More precisely, it will be proved that there exists an $N_0 = N_0(n)$ such that every $N \geq N_0$ can be represented in the form (1) when the number k of summands is of order $n \log n$, and that there exist infinitely many N that cannot be represented in the form (1) for $k \leq n$.

§1. The Circle Method for Waring's Problem

Let $J_{k,n}(N)$ denote the number of solutions in natural numbers $x_1, x_2, \ldots, x_k$ of the equation

$$x_1^n + x_2^n + \ldots + x_k^n = N.$$

We shall henceforth assume that the natural number N is greater than some fixed $N_0 = N_0(n) > 0$, which depends only on n, $n \geq 3$. As in the proof of Lemma 1 of Chapter X, we have the following formula that expresses $J_{k,n}(N)$ as the integral of a trigonometric sum:

$$J = J_{k,n}(N) = \int_0^1 S^k(\alpha) e^{-2\pi i \alpha N} \, d\alpha = \int_{-\kappa}^{1-\kappa} S^k(\alpha) e^{-2\pi i \alpha N} \, d\alpha,$$

where now

$$S(\alpha) = \sum_{0 < x \le P} e^{2\pi i \alpha x^n}, \quad P = N^{1/n}, \quad \tau = 2nP^{n-1}, \quad \chi\tau = 1.$$

By Lemma 2 of Chapter X, each α in the interval $[-\chi, 1 - \chi]$ can be represented in the form

$$\alpha = \frac{a}{q} + z, \quad 1 \le q \le \tau, \quad (a, q) = 1, \quad |z| \le \frac{1}{q\tau}.$$

We denote by E_1 the set of all α for which $q \le P^{0.25}$ in the above representation, and we denote by E_2 the set of all other α. The set E_1 consists of non-intersecting intervals $E(a, q)$ of the form

$$\frac{a}{q} - \frac{1}{q\tau} \le \alpha \le \frac{a}{q} + \frac{1}{q\tau}, \quad 0 \le a < q, \quad (a, q) = 1, \quad q = 1, 2, \dots, [p^{0.25}].$$

Denote by J_1 the integral over the set E_1 and by J_2 the integral over the set E_2. Then

$$J = J_1 + J_2.$$

The goal of this section is to obtain an asymptotic formula for J_1. First we shall find an upper bound for the modulus of the "complete" trigonometric sum $S(a, q)$ and the modulus of the trigonometric integral $\gamma(z)$,

$$S(a, q) = \sum_{x=1}^{q} e^{2\pi i \frac{ax^n}{q}}, \quad (a, q) = 1;$$

$$\gamma(z) = \int_0^1 e^{2\pi i z x^n} \, dx.$$

Lemma 1. *For $|S(a, q)|$ we have the following inequality;*

$$|S(a, q)| \le n^{n^6} q^{1 - 1/n}.$$

Proof. If $q = q_1 q_2$, where $(q_1, q_2) = 1$, then

$$S(a, q) = S(a_1 q_1) S(a_2 q_2).$$

Indeed, the expression $x_1 q_2 + x_2 q_1$ runs through a complete system or residues modulo q as x_1 and x_2 run through complete systems of residues modulo q_1 and q_2, respectively. Moreover,

$$(x_1 q_2 + x_2 q_1)^n \equiv x_1^n q_2^n + x_2^n q_1^n \pmod{q}.$$

Therefore,

$$S(a, q) = \sum_{x_1=1}^{q_1} \sum_{x_2=1}^{q_2} e^{2\pi i \frac{a(x_1 q_2 + x_2 q_1)^n}{q}}$$

$$= \sum_{x_1=1}^{q_1} e^{2\pi i \frac{a q_2^{n-1}}{q_1} x_1^n} \sum_{x_2=1}^{q_2} e^{2\pi i \frac{a q_1^{n-1}}{q_2} x_2^n} = S(a_1, q_1) S(a_2, q_2),$$

where

$$a_1 \equiv aq_2^{n-1}(\text{mod } q_1), (a_1, q_1) = 1, a_2 \equiv aq_1^{n-1}(\text{mod } q_2), (a_2, q_2) = 1.$$

It follows that

$$S(a, q) = S(a_1, p_1^{\alpha_1}) \dots S(a_r, p_r^{\alpha_r}), \tag{2}$$

where $q = p_1^{\alpha_1} \dots p_r^{\alpha_r}$ is the canonical decomposition of the number q. We shall estimate $|S(a, p^\alpha)|$, where $\alpha \geq 1$ and p is prime. Let $\alpha = 1$. Then

$$S(a, p) = \sum_{x=1}^{p} e^{2\pi i \frac{a}{p} x^n} = \frac{1}{p-1} \sum_{y=1}^{p-1} \sum_{x=1}^{p} e^{2\pi i \frac{ax^n y^n}{p}}.$$

Since the congruence $y^n \equiv \lambda \pmod p$, $1 \leq y \leq p-1$, has at most n solutions, we have

$$|S(a, p)|^2 \leq \frac{1}{p-1} \sum_{y=1}^{p-1} \left| \sum_{x=1}^{p} e^{2\pi i \frac{ax^n}{p} y^n} \right|^2$$

$$\leq \frac{n}{p-1} \sum_{\lambda=1}^{p-1} \left| \sum_{x=1}^{p} e^{2\pi i \frac{ax^n}{p} \lambda} \right|^2 = \frac{n}{p-1}(pK - p^2),$$

where K denotes the number of solutions of the congruence

$$x_1^n \equiv x_2^n (\text{mod } p), \quad 1 \leq x_1, x_2 \leq p.$$

Since $K \leq 1 + n(p-1)$, we have

$$|S(a, p)|^2 \leq \frac{n}{p-1}(p - p^2 + np(p-1)) < n^2 p, \quad |S(a, p)| < n\sqrt{p}.$$

Now let $1 < \alpha \leq n$, $(n, p) = 1$. Then

$$S(a, p^\alpha) = \sum_{y=0}^{p-1} \sum_{z=1}^{p^{\alpha-1}} e^{2\pi i \frac{a(yp^{\alpha-1} + z)^n}{p^\alpha}}$$

$$= \sum_{z=1}^{p^{\alpha-1}} e^{2\pi i \frac{az^n}{p^\alpha}} \sum_{y=0}^{p-1} e^{2\pi i \frac{anz^{n-1}y}{p}} = p \sum_{\substack{z=1 \\ z \equiv 0(\text{mod } p)}}^{p^{\alpha-1}} e^{2\pi i \frac{az^n}{p^\alpha}} = p^{\alpha-1}.$$

If $\alpha > n$, then, denoting by τ the exponent of p in the canonical decomposition of the number n, we have

$$S(a, p^\alpha) = \sum_{y=0}^{p^{\tau+1}-1} \sum_{z=1}^{p^{\alpha-\tau-1}} e^{2\pi i \frac{a(p^{\alpha-\tau-1}y + z)^n}{p^\alpha}}$$

$$= \sum_{z=1}^{p^{\alpha-\tau-1}} e^{2\pi i \frac{az^n}{p^\alpha}} \sum_{y=1}^{p^{\tau+1}-1} e^{2\pi i \frac{anz^{n-1}y}{p^{\tau+1}}} = p^{\tau+1} \sum_{\substack{z=1 \\ z \equiv 0(\text{mod } p)}}^{p^{\alpha-\tau-1}} e^{2\pi i \frac{az^n}{p^\alpha}}$$

$$= p^{\tau+1} \sum_{z=1}^{p^{\alpha-\tau-2}} e^{2\pi i \frac{az^n}{p^{\alpha-n}}} = p^{n-1} S(a, p^{\alpha-n}).$$

Let us introduce the function

$$T(a, q) = q^{-1+1/n} S(a, q).$$

From the previous estimate for $S(a, p^{\alpha})$ we find that, for $1 \le \alpha \le n$, $(p, n) = p$

$$|T(a, p^{\alpha})| = p^{-\alpha(1 - 1/n)}|S(a, p^{\alpha})| \le p^{\alpha/n} \le p \le n;$$

for $\alpha = 1$, $(p, n) = 1$

$$|T(a, p^{\alpha})| < p^{-1+1/n} n\sqrt{p} \le np^{-1/6};$$

and for $1 < \alpha \le n$, $(p, n) = 1$

$$|T(\alpha, p^{\alpha})| = p^{-\alpha\left(1 - \frac{1}{n}\right)} p^{\alpha - 1} = p^{\alpha/n - 1} \le 1.$$

Thus, for $1 \le \alpha \le n$,

$$|T(a, p^{\alpha})| \le \begin{cases} n, & \text{if } p \le n^6; \\ 1, & \text{if } p > n^6. \end{cases}$$

The last inequality holds for any α, since for $\alpha > n$

$$T(a, p^{\alpha}) = p^{-\alpha(1 - 1/n)} p^{n-1} S(a, p^{\alpha - n}) = T(a, p^{\alpha - n}).$$

We obtain from (2)

$$|S(a, q)| q^{-1+1/n} = |T(a_1, p_1^{\alpha_1})| \dots |T(a_r, p_r^{\alpha_r})| \le n^{n^6},$$

which is what we had to prove.   □

Corollary. *For $k \ge 2n + 1$, the "singular series" $\sigma = \sigma(N)$ for Waring's problem converges*:

$$\sigma = \sigma(N) \sum_{q=1}^{\infty} \sum_{\substack{0 \le a < q \\ (a, q) = 1}} \left(\frac{1}{q} S(a, q)\right)^k e^{-2\pi i \frac{aN}{q}}.$$

Lemma 2. *The following inequality holds for $|\gamma(z)|$*:

$$|\gamma(z)| \le \min(1, 2|z|^{-1/n}) = Z(z).$$

Proof. We shall assume that $z > 2^n$ and prove the second assertion of the Lemma, since the Lemma is trivial for $0 \le z \le 2^n$. A change of variable $zx^n = u$ in the integral gives

$$\int_0^1 e^{2\pi i z x^n} dx = \frac{1}{n} z^{-1/n} \int_0^z u^{-1+1/n} e^{2\pi i u} du$$

$$= \frac{1}{n} z^{-1/n} \left(\int_0^1 u^{-1+1/n} e^{2\pi i u} du + \int_1^z u^{-1+1/n} e^{2\pi i u} du \right).$$

The absolute value of the first integral does not exceed

$$\int_0^1 u^{-1+1/n} du = n.$$

Integrating the second integral by parts, we obtain

$$\int_1^z u^{-1+1/n} e^{2\pi iu} du = \frac{1}{2\pi i} u^{-1+1/n} e^{2\pi iu} \Big|_1^z + \frac{1}{2\pi i}\left(1 - \frac{1}{n}\right) \int_1^z u^{-2+1/n} e^{2\pi iu} du$$

In absolute value, this does not exceed

$$\frac{1}{\pi} + \frac{1}{2\pi}\left(1 - \frac{1}{n}\right) \int_1^z u^{-2+1/n} du < \frac{3}{2\pi} < n.$$

From this we obtain

$$|\gamma(z)| < 2z^{-1/n},$$

which is what we had to prove.

$\square$

Corollary. *For $k > n$ the "singular integral" $\gamma = \gamma(n, k)$ in Waring's problem converges:*

$$\gamma = \gamma(n, k) = \int_{-\infty}^{\infty} \left(\int_0^1 e^{2\pi izx^n} dx\right)^h e^{-2\pi iz} dz.$$

Theorem 1. *For $k \geq 2n + 1$, the following formula holds for J:*

$$J_1 = \sigma\gamma N^{\frac{h}{n}-1} + O(N^{\frac{h}{n}-1-\frac{1}{4n^2}}),$$

where

$$\sigma = \sigma(N) = \sum_{q=1}^{\infty} \sum_{\substack{0 \leq n < q \\ (a,q)=1}} \left(\frac{1}{q}S(a,q)\right)^k e^{-2\pi i \frac{aN}{q}},$$

$$\gamma = \gamma(n, k) = \int_{-\infty}^{+\infty} \left(\int_0^1 e^{2\pi izx^n} dx\right)^k e^{-2\pi iz} dz.$$

Proof. By definition of J_1 and by the properties of the set E_1, we have

$$J_1 = \sum_{q \leq p^{0.25}} \sum_{\substack{0 \leq a < q \\ (a,q)=p}} \int_{-1/q\tau}^{+1/q\tau} S^k\left(\frac{a}{q} + z\right) e^{-2\pi i\left(\frac{a}{q} + z\right)N} dz.$$

We shall transform $S(a/q + z)$. We represent x, $1 \leq x \leq P$, in the form $x = qt + s$, where $s = 1, 2, \ldots, q$, and for fixed s the integer t varies between the limits $1 - s/q \leq t \leq P - s/q$. We have

$$S\left(\frac{a}{q} + z\right) = \sum_{1 \leq x \leq P} e^{2\pi i\left(\frac{a}{q} + z\right)x^n} = \sum_{s=1}^q e^{2\pi i \frac{a}{q}s^n} \sum_{\frac{1-s}{q} \leq t \leq \frac{P-s}{q}} e^{2\pi iz(qt+s)^n}.$$

Since

$$\left| \frac{d}{dt} z(qt + s)^n \right| = |nzq(qt + s)^{n-1}| \leq 1/2,$$

then by the Corollary to Lemma 1 of Chapter I we have

$$\sum_{1-s/q \leq t \leq P-s/q} e^{2\pi i z(qt+s)^n} = \int_{(1-s)/q}^{(P-s)/q} e^{2\pi i z(qt+s)^n} dt + O(1)$$

$$= \frac{1}{q} \int_0^P e^{2\pi i z x^n} dx + O(1) = \frac{P}{q} \gamma(zN) + O(1).$$

Thus,

$$S\left(\frac{a}{q} + z \right) = \frac{P}{q} S(a, q)\gamma(zN) + O(q). \tag{3}$$

From Lemmas 1 and 2 we find that

$$S^k\left(\frac{a}{q} + z \right) e^{2\pi i \left(\frac{a}{q} + z \right) N} = P^k \gamma^k(zN) e^{-2\pi i z N} \left(\frac{1}{q} S(a, q) \right)^k e^{-2\pi i \frac{aN}{q}}.$$

$$+ O\left(P^{k-1} q^{-\frac{k-1}{n}+1} Z^{k-1}(zN) \right) + O(q^k)$$

$$J_1 = P^k V + O(R),$$

where

$$V = \sum_{q \leq P^{0.25}} \sum_{\substack{0 \leq a < q \\ (a,q)=1}} \left(\frac{1}{q} S(a, q) \right)^k e^{-2\pi i \frac{N}{q}} \int_{-1/q\tau}^{+1/q\tau} \gamma^k(zN) e^{-2\pi i z N} dz,$$

$$R = P^{k-0.75} \int_{-\kappa}^{+\kappa} |\gamma(zN)|^k dz + P^{\frac{k+1}{4}-n+1}$$

$$\leq 2P^{k-0.75} \left(\int_0^{2^n N - 1} dz + \int_{2^n N - 1}^{\kappa} 2^k (zN)^{-k/n} dz \right) + P^{k-n-1}$$

$$= O(P^{k-n-0.75}).$$

We shall transform V. First, we have

$$\int_{-1/q\tau}^{+1/q\tau} \gamma^k(zN) e^{-2\pi i z N} dz = \frac{1}{N} \int_{-N/q\tau}^{+N/q\tau} \gamma^k(z) e^{-2\pi i z} dz$$

$$= \frac{1}{N} \int_{-\infty}^{\infty} \gamma^k(z) e^{-2\pi i z} dz + O(R_1) = \frac{1}{N} \gamma + O(R_1),$$

where

$$R_1 \leq \frac{1}{N} \int_{N/q\tau}^{\infty} |\gamma(z)|^k dz \leq \frac{2^k}{N} \int_{N/q\tau}^{\infty} z^{-\frac{k}{n}} dz = O\left((q\tau)^{\frac{k}{n}-1} P^{-k} \right).$$

Consequently,

$$V = \frac{1}{N}\gamma \sum_{\substack{q \leq P^{0.25}}} \sum_{\substack{0 \leq a < q \\ (a,q)=1}} \left(\frac{1}{q}S(a,q)\right)^k e^{-2\pi i \frac{aN}{q}}$$

$$+ O(P^{-n-0.75}) = \frac{1}{N}\gamma\sigma + O(R_2) + O(P^{-n-0.75}),$$

where

$$R_2 \leq \frac{1}{N} \sum_{\substack{q > P^{0.25}}} \sum_{\substack{0 \leq a < q \\ (a,q)=1}} |S(a,q)|^k q^{-k} = O(P^{-n-1/4n}).$$

Finally, we find

$$V = \frac{1}{N}\gamma\sigma + O(P^{-n-1/4n});$$

$$J_1 = \gamma\sigma N^{\frac{k}{n}-1} + O(N^{\frac{k}{n}-1-\frac{1}{4n^2}})$$

This completes the proof. $\qquad\qquad\qquad\qquad\qquad\qquad\qquad\qquad\qquad\square$

We shall next investigate $\sigma = \sigma(N)$ and compute $\gamma = \gamma(n,k)$.

Lemma 3. *There exists a positive constant* $c = c(n,k) > 0$ *that depends only on* n *and* k *such that the singular series* $\sigma = \sigma(N)$ *of Theorem 1 is greater than* c *for* $k \geq 4n$, *i.e.* $\sigma > c > 0$.

Proof. The function

$$\Phi(q) = \sum_{\substack{(a,q)=1 \\ 0 \leq a < q}} \left(\frac{S(a,q)}{q}\right)^k e^{-2\pi i \frac{\sigma}{q}N}$$

is multiplicative. Indeed, if $q = q_1 q_2, (q_1, q_2) = 1$, and $a = a_1 q_2 + a_2 q_1$, then

$$S(a,q) = \sum_{x_1=1}^{q_1} \sum_{x_2=1}^{q_2} e^{2\pi i \frac{a(x_1 q_2 + x_2 q_1)^n}{q}}$$

$$= \sum_{x_1=1}^{q_1} e^{2\pi i \frac{a q_2^{n-1} x_1^n}{q_1}} \sum_{x_2=1}^{q_2} e^{2\pi i \frac{a q_1^{n-1} x_2^n}{q_2}}$$

$$= \sum_{x_1=1}^{q_1} e^{2\pi i \frac{a_1 x_1^n}{q_1}} \sum_{x_2=1}^{q_2} e^{2\pi i \frac{a_2 x_2^n}{q_2}};$$

$$\Phi(q) = \sum_{\substack{(a_1,q_1)=1 \\ 0 \leq a_1 < q_1}} \sum_{\substack{(a_2,q_2)=1 \\ 0 \leq a_2 < q_2}} \left(\frac{S(a_1 q_2 + a_2 q_1, q_1 q_2)}{q_1 q_2}\right)^k e^{-2\pi i \left(\frac{a_1 N}{q_1} + \frac{a_2 N}{q_2}\right)}$$

$$= \Phi(q_1)\Phi(q_2).$$

Furthermore, since

$$\Phi(q) \ll q^{\frac{h}{n}+1}. \qquad\qquad\qquad\qquad\qquad\qquad\qquad (4)$$

then

$$\prod_{p \le X} (1 + \Phi(p) + \Phi(p^2) + \ldots) = \sum_{q \le X} \Phi(q) + R(X),$$

where

$$R(X) \ll \sum_{q > X} |\Phi(q)| \ll X^{-\frac{h}{n}+2}.$$

Taking the limit as $X \to +\infty$ in the last equation, we obtain

$$\sigma(N) = \prod_p (1 + \Phi(p) + \Phi(p^2) + \ldots).$$

Note that $\Phi(p^r)$ is a real number for $r \ge 1$.

We shall use the upper bound (4) again:

$$\left| \sum_{r=1}^{\infty} \Phi(p^r) \right| \le c_1(k, n) \sum_{r=1}^{\infty} p^{-\left(\frac{k}{n}-1\right)r} \le c_2(k, n)p^{-3}.$$

Therefore, for $p > c_2(k, n)$

$$1 + \Phi(p) + \Phi(p^2) + \ldots > 1 - 1/p^2,$$

i.e.

$$\sigma(N) = \left(\prod_{p \le c_2(h,n)} (1 + \Phi(p) + \Phi(p^2) + \ldots) \right.$$

$$\times \prod_{p > c_2(h,n)} (1 + \Phi(p) + \Phi(p^2) + \ldots)$$

$$\ge \frac{6}{\pi^2} \prod_{p \le c_2(k,n)} (1 + \Phi(p) + \Phi(p^2) + \ldots).$$

It remains to prove that each parenthesis in the last product is greater than zero.

Denote by $T_k(p^m)$ the number of solutions of the congruence

$$x_1^n + x_2^n + \ldots + x_h^n \equiv N (\mathrm{mod}\, p^m). \tag{5}$$

Then we have

$$\Phi(p^r) = \sum_{\substack{a=1 \\ (a,p)=1}}^{p^r} \left(p^{-r} \sum_{x=1}^{p^r} e^{2\pi i \frac{ax^n}{p^r}} \right)^k e^{-2\pi i \frac{a}{p^r} N}$$

$$= p^{-rk} \sum_{a=1}^{p^r} \left(\sum_{x=1}^{p^r} e^{2\pi i \frac{ax^n}{p^r}} \right)^k e^{-2\pi i \frac{a}{p^r} N}$$

$$- p^{-rk+k} \sum_{x=1}^{p^{r-1}} \sum_{x=1}^{p^{r-1}} e^{2\pi i \frac{ax^n}{p^{r-1}}} \right)^k e^{-2\pi i \frac{a}{p^{r-1}} N}$$

$$= p^{-r(k-1)} T_k(p^r) - p^{-(r-1)(k-1)} T_k(p^{r-1});$$

$$1 + \sum_{r=1}^{m} \Phi(p^r) = p^{-m(k-1)} T_k(p^m). \tag{6}$$

We shall estimate $T_k(p^m)$ from below for sufficiently large m. We shall first consider $T_k(p)$, where

$$\gamma = \begin{cases} \tau + 1, & \text{if } p > 2, \quad n = p^\tau n_1, \quad (n_1, p) = 1; \\ \tau + 2, & \text{if } p = 2, \quad n = p^\tau n_1, \quad (n_1, p) = 1. \end{cases}$$

We shall prove that $T_k(p^\gamma) > 0$ for $k \geq 4n$, i.e. congruence (5) has solutions $x_1^{(0)}, \ldots, x_k^{(0)}$ such that at least one of the numbers $x_j^{(0)}$, $1 \leq j \leq k$, does not divide p. Without loss of generality, we can assume that $0 < N < p^\gamma$, $(N, p) = 1$, $k = 4n - 1$.

If $p = 2$, then $p^\gamma = 2^{\gamma+2} \leq 4n$, and the required solution will be the following choice of integers:

$$x_1 = \ldots = x_N = 1, x_{N+1} = \ldots = x_k = 0.$$

Let $p > 2$ and let g be a primitive root of the modulus p. If

$$N \equiv g^\alpha (\text{mod } p^\gamma), \quad N_1 \equiv g^\beta (\text{mod } p^\gamma), \quad \alpha \equiv \beta (\text{mod } n),$$

then the set of solutions of the congruence

$$x_1^n + \ldots + x_k^n \equiv N (\text{mod } p^\gamma) \tag{7}$$

coincides with set of solutions of the congruence

$$x_1^n + \ldots + x_k^n \equiv N_1 (\text{mod } p^\gamma),$$

since $\alpha = \beta + n\delta$,

$$(x_1 g^\delta)^n + \ldots + (x_k g^\delta)^n \equiv N g^{\delta n} \equiv N_1 (\text{mod } p^\gamma).$$

Denote by $k(N)$ the smallest k for which (7) has the needed solution, and let m be the number of different $k(N)$. It is clear that $m \leq n$. We partition the set of all N into m classes, such that numbers N_1 and N_2 are in the same class if $k(N_1) = k(N_2)$, and let $N_1, N_2, \ldots, N_m$ be the smallest representatives in each of these classes, arranged in order of increasing size. We shall prove that $k(N_r) \leq 2r - 1$, $r = 1, 2, \ldots, m$. For $r = 1$ we must have $N_1 = 1$, $k(N_1) = 1 \leq 2 \cdot 1 - 1$. If the inequality has been proved for $r = 1, 2, \ldots, h$, then, considering the two numbers $N_{h+1} - 1$ and $N_{h+1} - 2$, we see that one of these is not a multiple of p, is less than N_{h+1}, and, consequently, belongs to a class already considered, i.e. $k(N_{h+1}) \leq 2h - 1 + 2 = 2(h + 1) - 1$, which is what we had to prove. Therefore,

$$k(N) \leq k(N_m) \leq 2m - 1 \leq 2n - 1 < 4n.$$

Thus, congruence (7) has solutions $x_1^{(0)}, \ldots, x_k^{(0)}$ such that $(x_1^{(0)}, p) = 1$.

Next, we shall prove that if the congruence

$$y^n \equiv a (\text{mod } p^\gamma) \tag{8}$$

is solvable with $(y, p) = 1$, then for any $m > \gamma$ the congruence

$$x^n \equiv a (\text{mod } p^m) \tag{9}$$

is solvable. Let y_0 be a solution of congruence (8), $(y_0, p) = 1$, and let g be a primitive root of the modulus p^m if $p > 2$, and let $g = 5$ if $p = 2$. Choose a natural number b such that

$$g^b y_0^n \equiv a \pmod{p^m}.$$

Then

$$g^b \equiv 1 \pmod{p^\gamma}, \quad b = p^\gamma (p - 1) b_1.$$

For an arbitrary natural number r we consider the expression

$$b + rp^{m-1}(p - 1) = p^\gamma (p - 1)(b_1 + rp^{m-1-\tau}).$$

Since $n = p^\tau n_1$, where $(n_1, p) = 1$, we choose r such that $b_1 + rp^{m-1-\tau}$ is divisible by n_1. Then

$$b_1 + rp^{m-1-\tau} = n_1 h; \quad b + rp^{m-1}(p - 1) = nh(p - 1);$$

$$g^{b+rp^{m-1}(p-1)} \equiv g^b \pmod{p^m}; \quad g^{b+rp^{m-1}(p-1)} y_0^n \equiv a \pmod{p^m}$$

and $x_0 = y_0 g^{h(p-1)}$ is a solution of the congruence (9).

We proceed to estimate $T_k(p^m)$ from below. Consider the congruence

$$x_1^n + (x_2 + p^\gamma y_2)^n + \ldots + (x_k + p^\gamma y_k)^n \equiv N \pmod{p^\gamma},$$

$$1 \leq x_1, x_2, \ldots, x_k \leq p^\gamma, \quad 1 \leq y_2, \ldots, y_k \leq p^{m-\gamma}.$$

For $k \geq 4n$ it has solutions $x_1^{(0)}, x_2^{(0)}, \ldots, x_k^{(0)}$ such that $(x_1^{(0)}, p) = 1$, i.e. it has

$$p^{(k-1)(m-\gamma)}$$

solutions in integers $x_1^{(0)}, \ldots, x_k^{(0)}, y_2, \ldots, y_k$. But then the congruence

$$x_1^n \equiv N - (x_2^{(0)} + p^\gamma y_2)^n - \ldots - (x_h^{(0)} + p^\gamma y_k)^n \pmod{p^m}$$

is solvable with respect to x_1 for any integers $y_2, \ldots, y_k$, $1 \leq y_2, \ldots, y_k \leq p^{m-\gamma}$, i.e.

$$T_k(p^m) \geq p^{(k-1)(m-\gamma)}.$$

It follows from this and from (6) that

$$1 + \sum_{r=1}^{m} \Phi(p^r) = p^{-m(k-1)} T_k(p^m) \geq p^{-\gamma(k-1)};$$

$$1 + \sum_{r=1}^{\infty} \Phi(p^r) \geq p^{-\gamma(k-1)};$$

$$\prod_{p \leq c_2(k,n)} (1 + \Phi(p) + \Phi(p^2) + \ldots) \geq \prod_{p \leq c_2(k,n)} p^{-\gamma(k-n)} \geq c_3(k, n) > 0;$$

$$\sigma = \sigma(N) > c(k, n) > 0.$$

This completes the proof of the Lemma.   □

Lemma 4. *For $k \geq n + 1$ we have*

$$\gamma = \gamma(n, k) = \int\limits_{-\infty}^{+\infty} \left(\int\limits_0^1 e^{2\pi i z u^n} du \right)^k e^{-2\pi i z} dz = \frac{\left(\Gamma\left(1 + \dfrac{1}{n} \right) \right)^k}{\Gamma\left(\dfrac{k}{n} \right)}.$$

Proof. Let us consider the more general integral

$$g(x) = \int\limits_{-\infty}^{+\infty} \left(\int\limits_0^1 e^{2\pi i z u^n} du \right)^k e^{-2\pi i x z} dz, \quad 0 \leq x \leq 2.$$

Since

$$\left| \int\limits_0^1 e^{2\pi i z u^n} du \right| \ll \min\left(1, \frac{1}{|z|^{1/n}} \right),$$

then $g(x)$ converges absolutely for $k \geq n + 1$.

The function $g(x)$ is continuous on the interval $0 < x < 2$. Indeed,

$$|g(x + \varDelta x) - g(x)| < 4 \int\limits_0^{+\infty} \left| \int\limits_0^1 e^{2\pi i z u^n} du \right|^k |\sin \pi \varDelta x z| \, dz$$

$$< 4\pi |\varDelta x| \int\limits_0^{|\varDelta x|^{1/3}} z \, dz + 8 \cdot 2^k \int\limits_{|\varDelta x|^{-1/3}}^{\infty} z^{-k/n} dz \ll |\varDelta x|^{1/3n}.$$

Therefore, for $0 < c < 2$, the function $F(c)$ defined by

$$F(c) = \int\limits_0^c g(x) dx,$$

is differentiable. Furthermore, for $0 < c \leq 1$

$$F(c) = \int\limits_0^c g(x) dx = \int\limits_{-\infty}^{+\infty} \left(\int\limits_0^1 e^{2\pi i z u^n} du \right)^k \frac{1 - e^{-2\pi i z c}}{2\pi i z} dz$$

$$= \int\limits_0^1 \dots \int\limits_0^1 du_1 \dots du_k \int\limits_{-\infty}^{+\infty}$$

$$\times \frac{e^{2\pi i z (u_1^n + \dots + u_k^n)} - e^{2\pi i z (u_1^n + \dots + u_k^n - c)}}{2\pi i z} dz$$

$$= \frac{1}{\pi} \int\limits_0^1 \dots \int\limits_0^1 du_1 \dots du_k \int\limits_0^{\infty} \left(\frac{\sin 2\pi z \lambda}{z} - \frac{\sin 2\pi z (\lambda - c)}{z} \right) dz,$$

where

$$\lambda = u_1^n + \dots + u_k^n.$$

Since

$$\int\limits_0^{\infty} \frac{\sin \alpha x}{x} dx = \frac{\pi}{2} \operatorname{sign} \alpha,$$

then

$$F(c) = \frac{1}{2} \int_0^1 \cdots \int_0^1 (\text{sign } \lambda - \text{sign}(\lambda - c)) \, du_1 \ldots du_k = \int_{0 \leq \lambda \leq c} \cdots \int du_1 \ldots du_k.$$

Making the change of variables

$$u_1 = t_1^{1/n} c^{1/n}, \ldots, u_k = t_k^{1/n} c^{1/n}$$

in the integral, we obtain Dirichlet's integral (Theorem 6 of Chapter III):

$$F(c) = n^{-k} c^{k/n} \int_{\substack{0 \leq t_1 + \ldots + t_k \leq 1 \\ 0 \leq t_1, \ldots, t_k \leq 1}} \cdots \int t^{(1/n) - 1} \cdots t^{(1/n) - 1} \, dt_1 \ldots dt_k$$

$$= n^{-k} c^{k/n} \frac{\Gamma^k\left(\dfrac{1}{n}\right)}{\Gamma\left(\dfrac{k}{n} + 1\right)} = \frac{n}{k} c^{k/n} \frac{\left(\Gamma\left(1 + \dfrac{1}{n}\right)\right)^k}{\Gamma\left(\dfrac{k}{n}\right)}.$$

Differentiating $F(c)$, we find $g(c)$:

$$g(c) = c^{\frac{k}{n} - 1} \frac{\left(\Gamma\left(1 + \dfrac{1}{n}\right)\right)^k}{\Gamma(k/n)}.$$

For $c = 1$ we obtain the statement of the Lemma. $\square$

§2. An Estimate for Weyl Sums and the Asymptotic Formula for Waring's Problem

In order to obtain the asymptotic formula for $J_{k,n}(N)$, we need a nontrivial upper bound for $|J_2|$. For this we need to be able to estimate $|S(\alpha)|$ for $\alpha \in E_2$

Definition. A *Weyl sum* is a trigonometric sum of the form

$$S = S(\alpha_{n+1}, \ldots, \alpha_1) = \sum_{\alpha = 1}^{P} e^{2\pi i f(x)},$$

where $f(x) = \alpha_{n+1} x^{n+1} + \ldots + \alpha_1 x$, and the coefficients α_v are real numbers for $v = n + 1, n, \ldots, 1$.

Theorem 2. *Let*

$$\alpha_{n+1} = \frac{a}{q} + \frac{\theta}{q^2}, \quad (a, q) = 1, \quad |\theta| \leq 1, \quad P^{1/4} \leq q \leq P^{n+1-1/4}.$$

Let S be a Weyl sum. Then

$$|S| \leq C(n)P^{1-\frac{1}{800n^2 \log n}}.$$

Proof. We shall use the notation introduced in Lemma 1 of Chapter VI, as well as Lemmas 1, 3, 5, and 6 and Theorem 1 of Chapter VI. The plan of the proof of the Theorem is similar to the plan of the proof of Theorem 2 of Chapter VI.

First, for $Y = [P^{1-1/n^2}]$, we have

$$S = W + O(Y),$$

where $W = Y^{-1} \sum\limits_{y=1}^{Y} \sum\limits_{x=1}^{P} e^{2\pi i f(x+y)}$. We decompose $f(x+y)$ by powers of x:

$$f(x+y) = \alpha_{n+1}x^{n+1} + g_i(y)x^n + \ldots + g_n(y)x + g_0(y).$$

For each integer $k \geq 1$, we apply Lemmas 3 and 1 of Chapter VI and find that

$$|W|^{2k} \leq Y^{-1} \sum_{y=1}^{Y} \left| \sum_{x=1}^{P} e^{2\pi i(\alpha_{n+1}\alpha^{n+1}+g_1(y)x^n+\ldots)} \right|^{2k}$$

$$= Y^{-1} \sum_{y=1}^{Y} \sum_{\lambda_1,\ldots,\lambda_n,\lambda_{n+1}} J_{k,n+1}(\lambda_1,\ldots,\lambda_n,\lambda_{n+1})$$

$$\times e^{2\pi i(\alpha_{n+1}\lambda_{n+1}+g_1(y)\lambda_n+\ldots)}$$

$$\leq Y^{-1} \sum_{\lambda_1,\ldots,\lambda_n,\lambda_{n+1}} J_{k,n+1}(\lambda_1,\ldots,\lambda_n,\lambda_{n+1})$$

$$\times \left| \sum_{y=1}^{Y} e^{2\pi i(\alpha_{n+1}\lambda_{n+1}+g_1(y)\lambda_n+\ldots)} \right|$$

$$= Y^{-1} \sum_{\lambda_1,\ldots,\lambda_n} J_{k,n}(\lambda_1,\ldots,\lambda_n) \left| \sum_{y=1}^{Y} e^{2\pi i(g_1(y)\lambda_n+\ldots+g_n(y)\lambda_1)} \right|$$

$$\leq Y^{-1} \sqrt{\sum_{\lambda_1,\ldots,\lambda_n} J_{k,n}^2(\lambda_1,\ldots,\lambda_n)}$$

$$\times \sqrt{\sum_{\lambda_1,\ldots,\lambda_n} \left| \sum_{y=1}^{Y} e^{2\pi i(g_1(y)\lambda_n+\ldots+g_n(y)\lambda_1)} \right|^2}. \tag{10}$$

By Lemma 1 of Chapter VI,

$$\sum_{\lambda_1,\ldots,\lambda_n} J_{k,n}^2(\lambda_1,\ldots,\lambda_n) \leq J_{k,n}(0,\ldots,0)P^{2k}. \tag{11}$$

By Theorem 1 of Chapter VI, for $k \geq n\tau$

$$J_{k,n}(0,\ldots,0) \leq (n\tau)^{6n\tau}(2n)^{4n(n+1)\tau} P^{2k-\frac{n(n+1)}{2}\left(1-\left(1-\frac{1}{n}\right)^{\tau}\right)}. \tag{12}$$

Moreover,

$$g_1(y) = (n+1)\alpha_{n+1}y + \alpha_n, \quad |\lambda_v| < kP^v, \quad v = 1, \ldots, n.$$

Therefore, applying Lemma 4 of Chapter VI, we obtain

$$\sum_{\lambda_1,\ldots,\lambda_n} \left| \sum_{y=1}^{Y} e^{2\pi i(g_1(y)\lambda_n + \ldots + g_n(y)\lambda_1)} \right|^2$$

$$= \sum_{y,y_1=1}^{Y} \sum_{\lambda_1,\ldots,\lambda_n} e^{2\pi i(\lambda_n(n+1)\alpha_{n+1}(y-y_1) + \ldots + \lambda_1(g_n(y) - g_n(y_1)))}$$

$$\leq \sum_{y,y_1=1}^{Y} \min\left(2kP^n, \frac{1}{\|((n+1)(y-y_1)\alpha_{n+1})\|} \right) \sum_{\lambda_1,\ldots,\lambda_{n+1}} 1$$

$$\leq (2k)^{n-1} P^{\frac{n(n-1)}{2}} Y \sum_{y=1}^{(n+1)Y} \min\left(2kP^n, \frac{1}{\|(\alpha_{n+1}y + \beta)\|} \right). \tag{13}$$

Finally, applying Lemma 5 of Chapter VI, we can estimate the last sum:

$$\sum_{y=1}^{(n+1)Y} \min\left(2kP^n, \frac{1}{\|(\alpha_{n+1}y + \beta)\|} \right)$$

$$\leq 6\left(\frac{(n+1)Y}{q} + 1 \right)(2kP^n + q\log q) \leq 8knP^{n+0.75}\log P.$$

From $(10) - (13)$ for $\tau = [4n\log n] + 1$, we obtain

$$|W| \leq c_1(n)P^{1-\frac{1}{800n^2\log n}}; \quad |S| \leq c(n)P^{1-\frac{1}{800n^2\log n}},$$

which is what we had to prove.   □

Theorem 3. *For $k \geq cn^2 \log n$, the number $J_{k,n}(N)$ of representations of a natural number N in the form (1) satisfies the asymptotic formula*

$$J_{k,n}(N) = \gamma\sigma(N)N^{\frac{k}{n}-1} + O(N^{\frac{k}{n}-\frac{c_1}{n^2}-1}),$$

where

$$\gamma = \frac{\left(\Gamma\left(1 + \frac{1}{n}\right) \right)^k}{\Gamma\left(\frac{k}{n}\right)}, \quad \sigma(N) > c_0(n,k) > 0.$$

Proof. Applying the circle method (cf. §1), we have

$$J = J_{k,n}(N) = J_1 + J_2$$

where

$$J_1 = \int_{E_1} S^k(\alpha)e^{-2\pi i\alpha N}\,d\alpha, \quad J_2 = \int_{E_2} S^k(\alpha)e^{-2\pi i\alpha N}\,d\alpha.$$

From Theorem 1 and Lemmas 2 and 3, it follows that

$$J_1 = \gamma\sigma(N)N^{\frac{k}{n}-1} + O(N^{\frac{k}{n}-1-\frac{1}{4n^2}}).$$

From Theorem 2 for $\alpha \in E_2$

$$|S(\alpha)| \leq c_3(n)P^{1-\frac{1}{800n^2\log n}}.$$

From this, using the same notation as in Lemma 1 of Chapter VI and using Theorem 1 of Chapter VI, we find for $n\tau \leq k_1 < k/2$.

$$|J_2| \leq c_4(n,k)P^{(k-2k_1)}\left(1 - \frac{1}{800n^2\log n}\right)\int_0^1 |S(\alpha)|^{2k_1}\,d\alpha$$

$$= c_4(n,k)P^{(k-2k_1)}\left(1 - \frac{1}{800n^2\log n}\right)$$

$$\times \sum_{\lambda_1,\ldots,\lambda_{n-1}} J_{k_1,n}(\lambda_1,\ldots,\lambda_{n-1},0)$$

$$\leq (2k)^n c_4(n,k)P^{(k-2k_1)}\left(1 - \frac{1}{800n^2\log n}\right)P^{2k_1-n+\frac{n^2+n}{2}}\left(1-\frac{1}{n}\right)^\tau.$$

Now take

$$\tau = [4n\log n] + 1 \geq 4n\log n, \quad k_1 = n\tau, \quad k = 2k_1 + 800n^2.$$

This proves the Theorem.   □

§3. An Estimate for $G(n)$

It will be useful to make the following definition.

Definition. For $n \geq 3$ the function $G(n)$ is equal to the smallest integer k such that every integer $N \geq N_0(n)$ can be represented as a sum k natural numbers of the form x.

Theorem 4. $G(n)$ satisfies the inequality

$$n < G(n) \leq cn\log n.$$

Proof. We consider the sequence of numbers of the form $X = P^n + P^{n-2}$, where $P \geq P_0(n)$ is a natural number. Since $[X^{1/n}] = P$, if $k \leq n$, then the number of integers not exceeding X that can be represented as the sum of k natural numbers of the form x^n is not greater than

$$P^k \leq P^n < X = P^n + P^{n-2}.$$

This implies the first assertion of the Theorem. To prove the second assertion, we consider the equation

$$x_1^n + x_2^n + \ldots + x_k^n + u_1^n + \ldots + u_m^n + u_{m+1}^n + \ldots + u_{2m}^n = N, \quad (14)$$

where $x_1, x_2, \ldots, x_k, u_1, \ldots, u_{2m}$, are natural numbers, and, moreover,

$$P_1 = \tfrac{1}{4} N^{1/n} < u_1, \quad u_{m+1} < \tfrac{1}{2} N^{1/n} = 2P_1,$$

$$P_2 = \tfrac{1}{2} P_1^{1-1/n} < u_2, \quad u_{m+2} < P_1^{1-1/n} = 2P_2,$$

$$\cdots\cdots\cdots\cdots\cdots\cdots\cdots\cdots\cdots\cdots$$

$$P_m = \tfrac{1}{2} P_{m-1}^{1-1/n} < u_m, \quad u_{2m} < P_{m-1}^{1-1/n} = 2P_m.$$

First,

$$4^{-n}N = P_1^n \le u_1^n + \ldots + u_m^n + u_{m+1}^n + \ldots + u_{2m}^n \le 4(2P_1)^n = 2^{-n+2}N.$$

Furthermore, the equation

$$u_1^n + \ldots + u_m^n = u_{m+1}^n + \ldots + u_{2m}^n \quad (15)$$

has only solutions of the form $u_1 = u_{m+1}, u_2 = u_{m+2}, \ldots, u_m = u_{2m}$. Indeed, if, for example, $u_s \ne u_{m+s}$, $s < m$, and $u_1 = u_{m+1}, \ldots, u_{s-1} = u_{m+s-1}$, then

$$|u_s^n - u_{m+s}^n| > nP_s^{n-1},$$

$$|u_{s+1}^n + \ldots + u_m^n - u_{m+ws+1}^n - \ldots - u_{2m}^n| \le (2P_{s+1})^n = P_s^{n-1},$$

and equation (15) is impossible.

Let $I(N)$ denote the number of solutions of equation (14). Then

$$I(N) = \int_0^1 S^k(\alpha)\, T_1^2(\alpha) \ldots T_m^2(\alpha) e^{-2\pi i \alpha N}\, d\alpha,$$

where

$$S(\alpha) = \sum_{1 \le x \le P} e^{2\pi i \alpha x^n}, \quad P = N^{1/n},$$

$$T_1(\alpha) = \sum_{u_1} e^{2\pi i \alpha u_1^n},$$

$$\cdots\cdots\cdots\cdots$$

$$T_m(\alpha) = \sum_{u_m} e^{2\pi i \alpha u_m^n}.$$

Applying the definition of the sets E_1 and E_2 of §1, we have

$$I(N) = I_1(N) + I_2(N).$$

Let us estimate $I_2(N)$. By Theorem 2, for $\alpha \in E_2$

$$|S(\alpha)| \le c_3(n) P^{1 - \frac{1}{800 n^2 \log n}}, \quad P = N^{\frac{1}{n}}.$$

Therefore,

$$|I_2(N)| \leq c_4(n, k) P^k \left(1 - \frac{1}{800n^2\log n}\right) \int_0^1 |T_1(\alpha)|^2 \ldots |T_m(\alpha)|^2 \, d\alpha.$$

The last integral equals the number of solutions of equation (15), i.e. the number of choices of $u_1, \ldots, u_m$ that do not exceed

$$P_1 P_2 \ldots P_m \leq c_5(n, m) N^{1 - \left(1 - \frac{1}{n}\right)^m}.$$

Consequently,

$$|I_2(N)| \leq c_4(n, k) P_1 P_2 \ldots P_m N^{\frac{k}{n} - \frac{k}{800n^3\log n}}.$$

Let us estimate $I_1(N)$ from below. By the definition of $I_1(N)$,

$$I_1(N) = \sum_{u_1, u_{m+1}} \cdots \sum_{u_m, u_{2m}} \int_{\dot{E}_1} S^k(\alpha) e^{-2\pi i\alpha(N - u_1^n \ldots - u_{2m}^n)} d\alpha.$$

But the integral

$$\int_{\dot{E}_1} S^k(\alpha) e^{-2\pi i\alpha N_1} d\alpha$$

for

$$\left(1 - \frac{4}{2^n}\right) N \leq N_1 \leq \left(1 - \frac{1}{4^n}\right) N$$

and $k \geq 4n$ can be calculated from Theorem 1 (cf. also Lemmas 2 and 3):

$$\int_{\dot{E}_1} S^k(\alpha) e^{-2\pi i\alpha N_1} d\alpha = \gamma\sigma(N_1) N_1^{\frac{k}{n} - 1} + O\left(N_1^{\frac{k}{n} - 1 - \frac{1}{4n^2}}\right)$$

$$\geq c(k, n) N^{\frac{k}{n} - 1} - c_1(k, n) N^{\frac{k}{n} - 1 - \frac{1}{4n^2}}.$$

From this we get

$$I_1(N) \geq \sum_{u_1, u_{m+1}} \cdots \sum_{u_m, u_{2m}} \left(c(k, n) N^{\frac{k}{n} - 1} - c_1(k, n) N^{\frac{k}{n} - 1 - \frac{1}{4n^2}}\right)$$

$$\geq 2^{-2m}(P_1 P_2 \ldots P_m)^2 c(k, n) N^{\frac{k}{n} - 1}$$

$$- c_1(k, n)(P_1 P_2 \ldots P_m)^2 N^{\frac{k}{n} - 1 - \frac{1}{4n^2}}.$$

Since

$$P_1 P_2 \ldots P_m \geq c_6(n, m) N^{1 - \left(1 - \frac{1}{n}\right)^m},$$

then for $k = 4n$, $m = [c_0 n \log n]$, and $N \geq N_0(n)$ we have

$$I(N) = I_1(N) + I_2(N) > 0,$$

which is what we had to prove.   □

Exercises

1. Let $p \geq 3$ be a prime number, let $n_1, \ldots, n_k$ be fixed natural numbers, and let T be the number of solutions of the congruence

$$x_1^{n_1} \ldots + x_k^{n_k} \equiv \lambda \pmod{p}.$$

Prove that for $\lambda \not\equiv 0 \pmod{p}$,

$$T = p^{k-1} + O(p^{0.5(k-1)}).$$

2. Let $p \geq 3$ be a fixed prime number,

$$Q = p^\alpha, \quad P \leq Q, m = \frac{\log Q}{\log P} \leq \sqrt{n}, \quad Q \to +\infty.$$

Prove that for any N the congruence

$$x_1^n + \ldots + x_k^n \equiv N \pmod{Q}, \quad 1 \leq x_1, \ldots, x_k \leq P,$$

is solvable if $k \geq 30m$, and there exists an N such that this congruence has no solutions for $k < m$.

3. Let $1 \leq r \leq n$, let $p > n$ be a prime number, $1 \leq P \leq p^r$. Let T be the number of solutions of the system of congruences

$$\begin{cases} x_1 + \ldots + x_n \equiv y_1 + \ldots + y_n \pmod{p}, \\ x_2^2 + \ldots + x_n^2 \equiv y_1^2 + \ldots + y_n^2 \pmod{p^2}. \\ \cdots \cdots \cdots \cdots \cdots \cdots \cdots \cdots \cdots \cdots \cdots \cdots \\ x_1^n + \ldots + x_n^n \equiv y_1^n + \ldots + y_n^n \pmod{p^n}. \end{cases}$$

$$1 \leq x_1, \ldots, y_n \leq P, \quad x_i \not\equiv x_j \pmod{p}, \quad i \neq j,$$

Then

$$T \leq n! \, p^{\frac{r(r-1)}{2}} P^n.$$

4. Prove that for any natural number n and $x \geq (2n)^2$, the interval $[x, 2x]$ contains at least n different prime numbers.

5. Let $P > (4n^2)^n$ and let $p_1, \ldots, p_n$ be distinct prime numbers belonging to the interval $[P^{1/n}, 2P^{1/n}]$. Consider the system of equations

$$
\begin{cases}
x_1 + \ldots + x_k = y_1 + \ldots + y_k, \\
x_1^2 + \ldots + x_k^2 = y_1^2 + \ldots + y_k^2, \\
\quad\ldots\ldots\ldots\ldots\ldots\ldots\ldots\ldots \\
x_1^n + \ldots + x_k^n = y_1^n + \ldots + y_k^n,
\end{cases}
$$

$$
1 \le x_1, \ldots, \quad y_k \le P,
$$

and denote by J_2 the solutions of this system with the following property: For each $p_j, j = 1, \ldots, n$, there are at most $n - 1$ numbers among the integers $x_1, \ldots, x_k$, and among the integers $y_1, \ldots, y_k$, that are not congruent modulo p_j. Prove that

$$
J_2 \le n^{2kn} P^{k-1}.
$$

6. Let $k \ge n$, $P \ge 1$, and let $J = J(P; n, k)$ denote the number of solutions of the system of equations in exercise 5. Then there exists a number p lying in the interval $[P^{1/n}, 2P^{1/n}]$ such that

$$
J = J(P; n, k) \le 4k^{2n} p^{2k + \frac{n(n-5)}{2}} P^n J(P_1; n, k - n) + (2n)^{2kn} P^k,
$$

where $P_1 = Pp^{-1} + 1$.

7. Let τ, n, and k be natural numbers, $\tau \ge 1$, $n \ge 2$, $k \ge n\tau$, and $P \ge 1$. Then the following estimate holds for the number $J = J(P; n, k)$ of solutions of the system of equations of exercise 5:

$$
J = J(P; n, k) \le n^{2n\Delta(n)} 2^{\kappa} (8k)^{2n\tau} P^{2k - \Delta(n)},
$$

where

$$
\Delta(n) = \frac{n(n+1)}{2} - \frac{n^2}{2}\left(1 - \frac{1}{n}\right)^{\tau},
$$

$$
\kappa = n^2\tau + \frac{n(n+1)}{2}\tau - \frac{n^2(n-1)}{2}\left(1 - \left(1 - \frac{1}{n}\right)^{\tau}\right) < \frac{3(n+1)^2\tau}{2}.
$$

8. Let $n \ge 2$. We shall construct a polynomial $f(x)$ of degree n with integer coefficients such that

$$
f(x) \equiv 0 \pmod{2^n}, \quad \text{if } x \equiv 0 \pmod 2,
$$

$$
f(x) \equiv 1 \pmod{2^n}, \quad \text{if } x \equiv 1 \pmod 2.
$$

9. Consider a system of congruences of the following form (Hilbert-Kamke system):

$$
\begin{cases}
x_1^n + \ldots + x_k^n \equiv N_n, \\
x_1^{n-1} + \ldots + x_k^{n-1} \equiv N_{n-1} \pmod{2^n}, \\
\quad\ldots\ldots\ldots\ldots\ldots\ldots\ldots\ldots \\
x_1 + \ldots + x_k \equiv N_1.
\end{cases}
$$

Here, $x_1, \ldots, x_k$ are unknowns, and $N_n, \ldots, N_1$ are fixed integers. Prove that for the solvability of this system it is necessary that k be not less than the smallest nonnegative residue modulo 2^n of the number

$$a_n N_n + a_{n-1} N_{n-1} + \ldots + a_1 N_1,$$

where

$$f(x) = a_n x^n + a_{n-1} x^{n-1} + \ldots + a_1 x$$

is the polynomial constructed in the preceding exercise.

10. Let $n = 2^h$, where $h \geq 2$ is an integer. Prove that, if

$$x_1^n + \ldots + x_k^n \equiv 0 \pmod{4n},$$

$k < 4n$ and $x_1^n \equiv \ldots \equiv x_k^n \equiv 0 \pmod{4n}$, then

$$x_1 \equiv \ldots \equiv x_k \equiv 0 \pmod 2.$$

11. Let $96m \geq n \geq 1024$, and let $j_1, \ldots, j_m$ be integers satisfying the relation

$$\frac{3}{16} n < j_1 < j_2 < \ldots < j_m \leq \frac{1}{2} n.$$

Consider the system of congruences

$$\begin{cases} x_1^{2^{j_1}} + \ldots + x_k^{2^{j_1}} \equiv 0 \pmod{2^{2^{j_1}}}, \\ \cdots\cdots\cdots\cdots\cdots\cdots\cdots\cdots\cdots \\ x_1^{2^{j_m}} + \ldots + x_k^{2^{j_m}} \equiv 0 \pmod{2^{2^{j_m}}} \end{cases}$$

with the condition that there are odd numbers among the unknowns of this system. Then the solvability of the system implies that $k \geq 2^u$, $u = n/32$.

12. We say that the form $F = F(x_1, \ldots, x_k)$ in k variables with integer coefficients represents 0 only trivially modulo p if for some natural number M the congruence $F(x_1, \ldots, x_k) \equiv 0 \pmod{p^M}$ implies that $x_1 \equiv \ldots \equiv x_k \equiv 0 \pmod p$.

Prove that for any natural number r there exists $n_0 = n_0(r)$ such that for any $n \geq n_0$ there exists a form $F(x_1, \ldots, x_k)$ with integral coefficients of degree not exceeding n, where the number k of variables satisfies

$$k \geq 2^u, \quad u = \frac{n}{\underbrace{(\log_2 n)(\log_2 \log_2 n) \ldots (\log_2 \ldots \log_2 n)}_{r} \underbrace{(\log_2 \ldots \log_2^3 n)}_{r+1}},$$

and which represents zero only trivially modulo 2.

13. a) Let p be an odd prime number, $m \geq n/64(p-1)$, $n \geq (p-1)^8$, and let $j_1, \ldots, j_m$ be integers satisfying the inequality

$$\frac{3n}{8(p-1)} < j_1 < \ldots < j_m \leq \frac{n}{p-1}.$$

We consider the system of congruences

$$
\begin{cases}
x_1^{j_1(p-1)} + \ldots + x_k^{j_1(p-1)} \equiv 0 (\mathrm{mod}\ p^{j_1(p-1)}). \\
\ldots\ldots\ldots\ldots\ldots\ldots\ldots\ldots\ldots\ldots\ldots\ldots\ldots\ldots \\
x_1^{j_m(p-1)} + \ldots + x_k^{j_m(p-1)} \equiv 0 (\mathrm{mod}\ p^{j_m(p-1)})
\end{cases}
$$

with the condition that there is no multiple of p among the unknowns of this system. If the system is solvable, then $k \geq p^u$, where $u = n/64\ (p-1)$.

b) Let p be an odd prime number. For any natural number r there exists $n_1 = n_1(r; p)$ such that for $n \geq n_1$ there exists a form $F(x_1, \ldots, x_k)$ with integer coefficients and degree not exceeding n, where the number k of variables satisfies

$$
k \geq P^u, u = \frac{n}{\underbrace{(\log_p n)(\log_p \log_p n) \ldots (\log_p \ldots \log_p n)}_{r}\underbrace{(\log_p \ldots \log_p^3 n)}_{r+1}}
$$

and which represents zero only trivially modulo p.

14. Use the fact, that for $k \geq cn^2 \ln n$, we have the estimate (the "simplified upper bound for Vinogradov's integral")

$$
J = J_{k,n}(P) \leq c_1(n)P^{2k - \frac{n^2+n}{2}},
$$

to prove the following theorem:

Let $f(x) = a_{n+1}x^{n+1} + \ldots + \alpha_1 x$, where the α_ν's are real numbers, and

$$
\alpha_{n+1} = \frac{a}{q} + \frac{\theta}{q^2}, \quad 1 \leq q \leq P^{n+1}, \quad (a, q) = 1, \quad |\theta| \leq 1.
$$

Then

$$
\left| \sum_{\alpha \leq P} e^{2\pi i(x_{n+1}a^{n+1} + \ldots + \alpha_1 x)} \right| \leq c_2(n)P\Delta
$$

where

$$
\Delta = (\min(P, P^{n+1}q^{-1}, q))^{-\frac{1}{16cn^2 \ln n}}.
$$

15. a) Find an asymptotic formula for the number of solutions of the equation

$$
p_1 + p_2 + x^n = N,
$$

where $n \geq 2$ is fixed, p_1 and p_2 are prime numbers, and x is a natural number.

b) Find an asymptotic formula for the number of solutions of the equation

$$
p + x^2 + y^2 + z^n = N,
$$

where $n \geq 2$ is fixed, p is a prime number, and x, y, z are natural numbers.

Hints for the Solution of the Exercises

Chapter I

1. Repeat the proof of Theorem 4, and set $m = [10(b - a)D]$. Instead of the asymptotic formula for I_n, use the estimate $I_n \ll \sqrt{A}$.

2. α) In Lemma 1, let $r = 1$ and $\Delta = A^{1/3}$. Apply the following estimate to the coefficients of the Fourier series for the function $\psi(x)$:

$$|g(m)| \leq \begin{cases} \dfrac{1}{|m|}, & \text{if } 1 \leq |m| \leq A^{1/3}; \\[2ex] \dfrac{1}{|m|^2} A^{1/3}, & \text{if } |m| > A^{1/3}; \end{cases}$$

For $1 \leq |m| \leq A^{2/3}$, estimate the sum U_m

$$U_m = \sum_{a < x \leq b} e^{2\pi i m f(x)},$$

by using the result of exercise 1 (cf. also the proofs of Theorems 6 and 7).

β) The follows from α) and Theorem 1.

3. (V. Jarnik) Let $N \gg 1$, $m = \sum_{n=1}^{N} \varphi(n)$, and let $\xi_\nu = l_\nu / k_\nu$ be the Farey fractions of order N. Let $K_\nu = \sum_{r=1}^{\nu} k_r$, $L_\nu = \sum_{r=1}^{\nu} l_r$, and let M_ν be the point in the XY-plane with coordinates K_ν, L_ν, i.e. $M_\nu = (K_\nu, L_\nu)$. Draw a curve $y = f(x)$ through the point M such that

$$f''(x) \gg 1/N^3, \quad 1 = K_1 \leq x \leq K_m, \quad K_m \gg N^3,$$

$$0 < f'(x) \ll 1, \quad K_1 \leq x \leq K_m.$$

(Cf. V. Jarnik, Über die Gitterpunkte auf konvexen Kurven, Math. Zeit. 24 (1925), 500–518.)

4. The planes $x = 0$, $y = 0$, $z = 0$, $x = y$, $y = z$, and $z = x$ divide the sphere into 48 equal regions. Consider the following one of these regions:

$$0 \leq y \leq R/\sqrt{3}, \quad y \leq x \leq \sqrt{(R^2 - y^2)/2}, \quad x \leq z \leq \sqrt{R^2 - y^2 - x^2}$$

and use an argument similar to the proof of Theorem 2 (cf. Vinogradov [12]).

5. α) Apply the Corollary to Theorem 4.

β) Let $q = [a^{36/41}t^{-11/41}]$ and apply Lemma 3 to the sum. To the new trigonometric sum, apply Theorem 4, and to the next trigonometric sum apply Theorem 5 with $k = 5$ (cf. Titchmarsh [9]).

Chapter II

1. Let E be the set of points in the interval $[0, 1]$ for which $|f(x)| \leq A$. Then $\mu(E) = \mu$. Then one can find n points $x_1, x_2, \ldots, x_n$, in E such that $|x_k - x_{j'}| \geq |k - j| \mu/(n - 1)$. Consider the following system of equations:

$$f(x_i) = \theta_i A, \quad i = 1, 2, \ldots, n, \quad |\theta_i| \leq 1$$

This system is linear with respect to $\alpha_0, \alpha_1, \ldots, \alpha_{n-1}$. Now find α_j such that $|\alpha_j| = \alpha$.

2. Let

$$U = \int\limits_0^1 \cos 2\pi f(x)\,dx.$$

Divide the interval of integration into two sets E_1 and E_2. Let E_1 consist of all points for which

$$|f'(x)| \leq \left(\frac{n-1}{4e}\right)^{1-\frac{1}{n}} \alpha^{-\frac{1}{n}}$$

and let E_2 consist of the remaining points. The integral over E_1 can be estimated trivially by $\mu(E_1)$. The set E_2 can be decomposed into $\leq 2n - 2$ subintervals, in each of which $f'(x)$ is monotone and of constant sign. Now consider the integral over one of these subintervals. It will be useful here to apply the method of estimating an integral that was applied in the proof of Theorem 4 of Chapter I (cf. Vinogradov [10]).

3. The proof is by induction on the number of variables r. Represent the polynomial in the following form:

$$f(x_1, \ldots, x_r) = \sum_{t_1 = 0}^{n} \cdots \sum_{t_{r-1} = 0}^{n} x_1^{t_1} \ldots x_{r-1}^{t_{r-1}} \varphi(x_r).$$

and then apply the results of exercises 1 and 2. (Cf. V.N. Chubarikov, Multiple rational trigonometric sums and multiple integrals, Mat. Zametki 20 (1976), no. 1, 61–68.)

4. First prove that

$$J = \frac{(-1)^{r-1}}{(r-1)!} \int\limits_0^1 e^{2\pi i a x^n} (\ln x)^{r-1}\,dx.$$

5. As in exercise 1, choose n points $x_1, x_2, \ldots, x_n$ belonging to U such that

$$|x_k - x_j| \geq |k - j| \mu/(n - 1), \quad \mu = \mu(U),$$

Then construct the Lagrange interpolation polynomial $g(x)$ corresponding to $f'(x)$ and the interpolation points $x_1, x_2, \ldots, x_n$,

$$g(x) = \sum_{v=1}^{n} f'(x_v) \frac{(x - x_1) \ldots (x - x_{v-1})(x - x_{v+1}) \ldots (x - x_n)}{(x_v - x_1) \ldots (x_v - x_{v-1})(x_v - x_{v+1}) \ldots (x_v - x_n)}$$

and apply Rolle's theorem $n - 1$ times consecutively to the function $F(x) = g(x) - f'(x)$. (Cf. G.I. Arkhipov, A.A. Karatsuba, and V.N. Chubarikov, Trigonometric integrals, Izv. Akad. Nauk SSSR Ser. Mat. 43 (1979), no. 5, 971–1003, 1197.)

6. This is similar to exercise 2 and uses exercise 5.

7. The covering is constructed in $\leq n$ steps, by considering the sequence of functions

$$\beta_{n-k}(x) = \frac{1}{(n-k)!} f^{(n-k)}(x)$$

for $k = 0, 1, \ldots, n - 1$, and noting that, for any $D > 0$, the number of intervals in which each point satisfies the inequality

$$|\beta_{n-k}(x)| < D,$$

does not exceed k, and the number of intervals in which each point satisfies the inequality

$$|\beta_{n-k}(x)| \geq D.$$

does not exceed $k + 1$.

8. This follows from exercises 6 and 7.

9. Apply the result of exercise 8, after first computing an upper bound for the volume of the domain $\Omega = \Omega(\alpha_n, \ldots, \alpha_1)$ of all points $\alpha_n, \ldots, \alpha_1$ for which the value of H does not exceed some natural number P. To do this, for $r = 1, 2, \ldots, P$, consider the domain $\Omega_r = \Omega_r(\alpha_n, \ldots, \alpha_1)$ of all points $\alpha_n, \ldots, \alpha_1$ which satisfy the inequality

$$\left| \beta_s \left(\frac{r}{P} \right) \right| \leq 2^n P^s, \quad s = 1, 2, \ldots, n.$$

Prove that

$$\mu(\Omega_r) = \int \ldots \int_{\Omega_r} d\alpha_n \ldots d\alpha_1 = 2^{n^2 + n} P^{(n^2 + n)/2} .$$

Show that each point of the domain Ω belongs to the domain Ω_r for some $r = 1, \ldots, P$. (For a proof of the convergence of for $2k \leq 0.5(n^2 + n) + 1$, and for a generalization of exercises 5–9, see the article by Arkhipov et al., referred to in exercise 5.)

Chapter III

1. See the proof of Theorem 4.1.
2. This follows from exercise 1.

3. This follows from exercise 2.

4. Use exercises 2 and 3 and the fact that the sum $S = 1$ when $N = 1$. (See Davenport [4] in connection with exercises 1–4.)

5. If n_1 and m_1 are the remainders when n and m, respectively, are divided by p, where $0 < n_1 < p$ and $0 < m_1 < p$, then the condition implies that $n_1 m_1 = kp$. Any prime divisor of $n_1 m_1$ must be less than p, and so, by the induction hypothesis, must divide k. Cancelling these prime divisors successively leads to the contradiction that $1 = k_1 p$.

6. This follows from exercise 5.

7. a) Let $0 < a_q = \max_{0 \leq j \leq k} |a_j|$, $p^u \| m + q$, $p^v \| k!$. Then $u \leq v$. (Cf. E.M. Nikishin, Logarithms of natural numbers, Izv. Akad. Nauk SSSR Ser. Mat., 43 (1979), no. 6, 1319–1327; Correction: "Logarithms of natural numbers," Izv. Akad. Nauk SSSR Ser. Mat., 44 (1980), no. 4, 972.)

b) (V.K. Ryzhov). This follows from the identity

$$((n + 2)(n - 1)^2)((n - 2)(n + 1)^2)(n^3) = ((n + 2)(n - 2)n)((n - 1)(n + 1)n)^2.$$

8. a) For $k = 1$, by definition, $\tau_k(n) \equiv 1$. Therefore,

$$\sum_{n \leq X} \tau_1(n) \leq X.$$

Assume that the statement holds for $k = m$. We shall prove it for $k = m + 1$. Since $\tau_{k+1}(n) = \sum_{d \backslash n} \tau_k(d)$, then

$$\sum_{n \leq X} \tau_{k+1}(n) = \sum_{n \leq X} \sum_{d \backslash n} \tau_k(d) = \sum_{d \leq X} \tau_k(d) \sum_{\substack{n \equiv 0 \,(\mathrm{mod}\, d) \\ n \leq X}} 1$$

$$\leq X \sum_{d \leq X} \frac{\tau_k(d)}{d} = X \left(\int_1^X \left(\sum_{d \leq u} \tau_k(d) \right) u^{-2} du + X^{-1} \sum_{d \leq X} \tau_k(d) \right)$$

$$< X \left(\int_1^X \frac{1}{(k-1)!} u^{-1} (\ln u + k - 1)^{k-1} du + \frac{1}{(k-1)!} (\ln X + k - 1)^{k-1} \right)$$

$$\leq X \left(\frac{1}{k!} (\ln X + k - 1)^k + \frac{1}{(k-1)!} (\ln X + k - 1)^{k-1} \right)$$

$$\leq \frac{1}{k!} X (\ln X + k)^k,$$

which is what we had to prove.

b) Just as in part a), the inequality will be proved by induction. It is trivial for $k = 1$. Assume that the inequality holds for $k = m$. We shall prove it for $k = m + 1$. Use the Abel transformation, the inequality $\tau_k(nr) \leq \tau_k(n)\tau_k(r)$, and the reasoning used in a). (Cf. K.K. Mardzhanishvili, An estimate for an arithmetic sum, Doklady Akad. Nauk SSSR 22 (1939), no. 7, 391–393.)

9. Let $F(m, n)$ be an arithmetic function. Then

$$\sum_{n\leq N} F(1, n) = \sum_{m\leq u} \sum_{n\leq Nm^{-1}} F(m, n) \sum_{d\backslash m} \mu(d)$$

$$= \sum_{d\leq u} \sum_{r\leq ud^{-1}} \sum_{n\leq N(dr)^{-1}} \mu(d)F(dr, n)$$

$$= \sum_{d\leq u} \sum_{r\leq Nd^{-1}} \sum_{n\leq N(dr)^{-1}} \mu(d)F(dr, n)$$

$$- \sum_{d\leq u} \sum_{ud^{-1}<r\leq Nd^{-1}} \sum_{n\leq N(dr)^{-1}} \mu(d)F(dr, n).$$

Setting $m = dr$, we can rewrite the last multiple sum as:

$$\sum_{u<m\leq N} \sum_{\substack{d\backslash m \\ d\leq u}} \mu(d) \sum_{n\leq Nm^{-1}} F(m, n).$$

If we take

$$F(m, n) = \begin{cases} \Lambda(n)f(nm), & u < n; \\ 0, & u \geq n, \end{cases}$$

then we obtain

$$\sum_{u<n\leq N} \Lambda(n)f(n) = \sum_{d\leq u} \sum_{r\leq Nd^{-1}} \sum_{u<n\leq N(dr)^{-1}} \mu(d)\Lambda(n)f(n\,dr)$$

$$- \sum_{u<m\leq N} \sum_{n\leq Nm^{-1}} \left(\sum_{\substack{d\backslash m \\ d\leq u}} \mu(d) \right) \Lambda(n)f(nm) = S_1 - S_2,$$

$$S_1 = \sum_{d\leq u} \sum_{rd\leq N} \sum_{ndr\leq N} \mu(d)\Lambda(n)f(n\,dr) - \sum_{d<u} \sum_{rd\leq N} \sum_{\substack{n\leq u \\ nrd\leq N}} \mu(d)\Lambda(n)f(n\,dr)$$

$$= \sum_{d\leq u} \mu(d) \sum_{l\leq Nd^{-1}} (\log l)f(ld) - \sum_{d\leq u} \mu(d) \sum_{n\leq u} \Lambda(n) \sum_{r\leq N(dn)^{-1}} f(n\,dr),$$

$$S_2 = \sum_{u<m\leq Nu^{-1}} \left(\sum_{\substack{d\backslash m \\ d\leq u}} \mu(d) \right) \sum_{u<n\leq Nm^{-1}} \Lambda(n)f(nm).$$

which is what we need. (Cf. R.C. Vaughan, On the distribution of αp modulo 1, Mathematika 24 (1977), no. 2, 135–141.)

Chapter IV

1. For $\mathrm{Re}\,\tau > 0$ and $\mathrm{Re}\,s > 0$, make a change in the variable of integration $x \to x\tau$, which rotates the line of integration by the angle $\varphi = \arg\tau$. Applying

Corollary 1 of Lemma 3, we have (compare the proof of Theorem 1):

$$\pi^{-\frac{s}{2}}\Gamma\left(\frac{s}{2}\right)\zeta(s) = \int_0^\infty x^{\frac{s}{2}-1}\left(\sum_{n=1}^\infty e^{-\pi n^2 x}\right) dx = \int_0^\infty (\tau x)^{\frac{s}{2}-1}\left(\sum_{n=1}^\infty e^{-\pi n^2 \tau x}\right) d(\tau x)$$

$$= \tau^{\frac{s}{2}}\int_1^\infty x^{\frac{s}{2}-1}\left(\sum_{n=1}^\infty e^{-\pi n^2 \tau x}\right) dx + \tau^{\frac{s}{2}}\int_0^\infty x^{-\frac{s}{2}-1}\left(\sum_{n=1}^\infty e^{-\pi n^2 \tau \frac{1}{x}}\right) dx$$

$$= \tau^{\frac{s}{2}}\int_1^\infty x^{\frac{s}{2}-1}\left(\sum_{n=1}^\infty e^{-\pi n^2 \tau x}\right) dx - \frac{\tau^{\frac{s}{2}}}{s} - \frac{\tau^{\frac{s}{2}-1}}{1-s}$$

$$+ \tau^{\frac{s}{2}-1}\int_1^\infty x^{-\frac{s}{2}-\frac{1}{2}}\left(\sum_{n=1}^\infty e^{-\pi n^2 \frac{1}{\tau}x}\right) dx.$$

Moreover,

$$\int_1^\infty x^{\frac{s}{2}-1}e^{-\pi n^2 \tau x}dx = \pi^{-\frac{s}{2}}n^{-s}\tau^{-\frac{s}{2}}\Gamma\left(\frac{s}{2}, \pi n^2 \tau\right);$$

$$\int_1^\infty x^{-\frac{s}{2}-\frac{1}{2}}e^{-\pi n^2 \frac{1}{\tau}x}dx = \tau^{-\frac{s}{2}+\frac{1}{2}}\pi^{\frac{s}{2}-\frac{1}{2}}n^{s-1}\Gamma\left(\frac{1-s}{2}, \pi n^2 \frac{1}{\tau}\right).$$

The exercise follows from this.

2. a) Let $\tau = 1$ and repeat the argument used in the proof of Theorem 1.

b). Let $\tau = \dfrac{t}{\pi X^2}e^{i\left(\frac{\pi}{2}-\frac{1}{t}\right)}$, $t > 0$, estimate the corresponding integral, and apply Lemma 2 of Chapter I. (Cf. A.F. Lavrik, The approximate functional equation for Dirichlet's function, Izv. Akad. Nauk SSSR Ser. Mat. 32 (1968), no. 1, 134–185.)

3. It follows from the functional equation for the zeta function (Theorem 1) that for $s = 1/2 + it$

$$\pi^{-\frac{it}{2}}\Gamma\left(\frac{1}{4}+it\right)\zeta\left(\frac{1}{2}+it\right) = \pi^{\frac{it}{2}}\Gamma\left(\frac{1}{4}-\frac{it}{2}\right)\zeta\left(\frac{1}{2}-it\right),$$

i.e.

$$Z(t) = \overline{Z(t)}.$$

4. This follows from the identity

$$e^{i2\theta(t)} = \frac{\pi^{-it}\Gamma\left(\dfrac{1}{4}+\dfrac{it}{2}\right)}{\Gamma\left(\dfrac{1}{4}-\dfrac{it}{2}\right)}$$

and from the formula

$$\ln\Gamma(s) = \left(s-\frac{1}{2}\right)\ln s - s + \ln\sqrt{2\pi} + \int_0^\infty \frac{\rho(u)\,du}{u+s}.$$

5. By Leibniz's formula for integral $k \geq 0$,

$$Z^{(k)}(t) = \sum_{r=0}^{k} \binom{k}{r} \frac{d^r}{dt^r} (e^{i\theta(t)}) \frac{d^{k-r}}{dt^{k-r}} \left(\zeta \left(\frac{1}{2} + it \right) \right). \tag{1}$$

From exercise 4

$$\frac{d^r}{dt^r} (e^{i\theta(t)}) = i^r (\theta'(t))^r e^{i\theta(t)} + O(t^{-1} \ln^{r-1} t). \tag{2}$$

From Lemma 2 for $0 \leq m \leq k$ and $N \geq 1$,

$$\zeta^{(m)} \left(\frac{1}{2} + it \right) = (-i)^m \sum_{n=1}^{N} \frac{(\ln n)^m}{n^{\frac{1}{2}+it}} + \frac{d^m}{dt^m} \left(\frac{N^{\frac{1}{2}-it}}{it - \frac{1}{2}} \right) + O\left(\frac{t \ln^m N}{\sqrt{N}} \right).$$

Setting $N > N_0 = t/2\pi$, we transform the sum S,

$$S = \sum_{N_0 < n \leq N} \frac{(\ln n)^m}{n^{\frac{1}{2}+it}},$$

by applying the Corollary to Lemma 1 of Chapter I (see the proof of Theorem 6). Letting $N \to +\infty$, we obtain

$$\zeta^{(m)} \left(\frac{1}{2} + it \right) = (-i)^m \sum_{n \leq t/2\pi} \frac{(\ln n)^m}{n^{\frac{1}{2}+it}} + O\left(\frac{(\ln t)^m}{\sqrt{t}} \right).$$

Apply Theorem 2 of Chapter I to prove the formula

$$\sum_{\sqrt{\frac{t}{2\pi}} < n \leq \frac{t}{2\pi}} \frac{(\ln n)^m}{\sqrt{n}} n^{-it} = e^{i\theta_1(t)} \sum_{n \leq \sqrt{\frac{t}{2\pi}}} \frac{\left(\ln \frac{t}{2\pi n} \right)^m}{n^{\frac{1}{2}-it}} + O(t^{-\frac{1}{4}} (\ln t)^{m+1}),$$

where

$$\theta_1(t) = -t \ln t + t \ln 2\pi + t + \pi/4.$$

The exercise now follows from formulas (1) and (2) and the approximate functional equation for $\zeta^{(m)}(1/2 + it)$:

$$\zeta^{(m)} \left(\frac{1}{2} + it \right) = (-t)^m \left(\sum_{n \leq \sqrt{\frac{t}{2\pi}}} \frac{(\ln n)^m}{n^{\frac{1}{2}+it}} + e^{i\theta_1(t)} \sum_{n \leq \sqrt{\frac{t}{2\pi}}} \frac{\left(\ln \frac{t}{2\pi n} \right)^m}{n^{\frac{1}{2}-it}} \right)$$

$$+ O(t^{-\frac{1}{4}} (\ln t)^{m+1}).$$

6. Let $\Delta = 8H_1^{-1} \ln P$ and split the sum S_1 into two parts:

$$S_1 = S_{11} + S_{12},$$

where

$$S_{11} = \sum_{n \le (1-\varDelta)P}, \quad S_{12} = \sum_{P \ge n > (1-\varDelta)P}.$$

For S_{11} we have

$$S_{11} \ll \sum_{n \le (1-\varDelta)P} \frac{\left(\ln \dfrac{P}{n}\right)^k}{\sqrt{n}} \left| \sum_{v=0}^{H_1-1} e^{2\pi i v \alpha} \right|^r ; \quad \left| \sum_{v=0}^{H_1-1} e^{2\pi i v \alpha} \right| \le \min\left(H_1, \frac{1}{\|\alpha\|}\right) \le \frac{H_1}{4}.$$

For S_{12} we have

$$S_{12} \ll H_1^r \left| \sum_{P(1-\varDelta) < n \le P} \frac{\left(\ln \dfrac{P}{n}\right)^k}{\sqrt{n}} e^{i \frac{\pi v}{\ln P} \ln n} \right|,$$

where v is an integer that satisfies the condition

$$T < \pi v / \ln P < T + H.$$

After a change of variables, we are led to the sums σ and $\sigma(a)$:

$$\sigma = \sum_{m < \varDelta P} \frac{\left(-\ln\left(1 - \dfrac{m}{p}\right)\right)^k}{\sqrt{P-m}} e^{it \ln(P-m)},$$

$$\sigma(a) = \sum_{a < m \le 2a} \frac{\left(-\ln\left(1 - \dfrac{m}{P}\right)\right)^k}{\sqrt{P-m}} e^{it \ln(P-m)}.$$

Finally, partial summation gives

$$\sigma(a) \ll a^k P^{-k-1/2} \left| \sum_{a < m \le a_1} e^{it \ln(P-m)} \right|, \quad a_1 \le 2a.$$

We estimate the last sum trivially if $a \le \sqrt[3]{P}$, and by the third derivative if $a > \sqrt[3]{P}$. For $|S_1|$ we obtain

$$|S_1| \ll H_1^r (\ln P)^k (T^{1/6} H^{-k-1} \ln^2 T + T^{-1/4} \ln T).$$

In $|S_2|$ we separate out the summand for $n = 1$, and we estimate the remaining terms in the same way that we estimated S_{11}. We obtain

$$S_2 = H_1^r (\ln P)^k \{1 + Q(T^{1/4}(4H_1^{-1} \ln P)^r) + O(T^{-1/4} \ln T)\}.$$

7. This follows from exercises 5 and 6. (Cf. J. Moser, A certain Hardy-Littlewood theorem in the theory of the Riemann zeta function, Acta Arith. 31 (1976), 45–51; A.A. Karatsuba, On the distance between consecutive zeros of the Riemann zeta function that lie on the critical line, Trudy Mat. Inst. Steklov, 157 (1981), 49–63, 235.)

Chapter V

1. For Re $s > 1$ we have

$$\frac{f(s)}{s} = \int_1^\infty A(\xi) \frac{d\xi}{\xi^{s+1}}.$$

Multiply both sides of this equation by $x^{s+1}/s + 1$, integrate over the interval $[b + iT, b - iT]$, and apply the reasoning used in Theorem 1. (Cf. A.A. Karatsuba, A uniform estimate for the remainder term in Dirichlet's divisor problem, Izv. Akad. Nauk SSSR Ser. Mat. 36 (1972), 474–483.)

2. Applying the corollary to Theorem 6 of Chapter IV, we obtain

$$\int_T^{2T} \left| \zeta\left(\frac{1}{2} + it\right) \right|^4 dt \ll \int_T^{2T} \left| \sum_{n \leq T} \frac{\tau'(n)}{\sqrt{n}} n^{it} \right|^2 dt + 1,$$

where $0 \leq \tau'(n) \leq \tau(n)$. We estimate the last integral by using the estimate

$$\sum_{n \leq X} \tau^2(n) \ll X \ln^3 X.$$

together with an argument similar to that used in the proof of Theorem 1.

3. (Hardy-Littlewood) For Re $s > 1$,

$$\zeta^4(s) = \sum_{n=1}^\infty \frac{\tau_4(n)}{n^s}.$$

Now apply Theorem 1 and exercise 2. (Cf. G.H. Hardy and J.E. Littlewood, The approximate functional equation in the theory of the zeta-function, with applications to the divisor problems of Dirichlet and Piltz, Proc. London Math. Soc. (2) 21 (1922), 39–74.)

4. It follows from the Corollary to Theorem 5 of Chapter IV that, for some $c > 0$ in the domain $2 \leq |t| \leq T$, $\sigma \geq 1 - c/\log T$, we have the estimate

$$\frac{\zeta'(s)}{\zeta(s)} = O(\log^2 T).$$

In the same region,

$$\int_\sigma^{1+1/\log T} d(\ln \zeta(u + i|t|)) = \ln \zeta\left(1 + \frac{1}{\log T} + i|t|\right) - \ln \zeta(\sigma + i|t|),$$

i.e.

$$|\ln \zeta(\sigma + i|t|)| \ll \ln \ln T, \quad \frac{1}{|\zeta(\sigma + i|t|)|} \ll \ln^A |t|.$$

Moreover, for $b > 1$ by Theorem 1,

$$\sum_{n \leq X} \mu(n) = \frac{1}{2\pi i} \int_{b-iT}^{b+iT} \frac{1}{\zeta(s)} \cdot \frac{X^s}{s} ds + O\left(\frac{X^b}{T(b-1)}\right) + O\left(\frac{X \ln X}{T}\right).$$

Then repeat the argument used in Theorem 2.

5. Sufficiency. First, a) and b) are equivalent by partial summation (cf. Theorem 2). Next, we obtain by partial summation for Re $s > 1$,

$$-\frac{\zeta'(s)}{\zeta(s)} = \sum_{n=1}^{\infty} \frac{\Lambda(n)}{n^s} = s \int_1^{\infty} \psi(x) x^{-s-1} dx = s \int_1^{\infty} x^{-s} dx + s \int_1^{\infty} R(x) x^{-s-1} dx$$

$$= \frac{s}{s-1} + s \int_1^{\infty} R(x) x^{-s-1} dx,$$

where $R(x) = O(x^{\gamma+\varepsilon})$. The last integral defines an analytic function in the half plane Re $s > \gamma$. Consequently, by the principle of analytic continuation, the function $\zeta'(s)/\zeta(s)$ is analytic everywhere in the half plane Re $s > \gamma$ except at the point $s = 1$, where it has a pole of order 1. It follows from this that $\zeta(s)$ has no zeros for Re $s > \gamma$. Conversely, if there are no zeroes in the half plane Re $s > \gamma$, then a) and b) follow from Theorem 3 (cf. the Corollary). Assertion c) follows from the same kind of argument used in exercise 4.

6. (J.E. Littlewood) Use Theorem 3 together with the Mellin transform

$$e^{-x} = \frac{1}{2\pi i} \int_{\alpha-i\infty}^{\alpha+i\infty} x^{-w} \Gamma(w) dw \quad (\text{Re } x > 0, \alpha > 0).$$

7. Prove that for $c > 0$ and $T \geq 2$,

$$\frac{1}{2\pi i} \int_{e-iT}^{e+iT} \frac{ds}{s} = \frac{1}{2} + O\left(\frac{1}{T}\right)$$

Then apply formula (2) of Chapter V, with Theorem 3 of Chapter IV and the method of proof used in Theorem 3.

8. a) For Re $s > 1$, consider the function $f(s)$ defined by

$$f(s) = \sum_{n=1}^{\infty} \frac{r(n)}{n^s} = \prod_p \left(1 + \frac{1}{\varphi^s(p)} + \frac{1}{\varphi^s(p^2)} + \ldots\right),$$

and compare it with $\zeta(s)$.

b) For Re $s > 1$, consider the function $\Phi(s)$ defined by

$$\Phi(s) = \sum_{n=1}^{\infty} \frac{1}{n^s \varphi(n)} = \prod_p \left(1 + \frac{1}{p^s \varphi(p)} + \frac{1}{p^{2s} \varphi(p^2)} + \ldots\right).$$

9. This follows from Theorem 2.

Remark. In exercises 8 and 9 we obtain a remainder term different from the remainder term in Theorem 2 (cf. Theorem 3 of Chapter VI).

Chapter VI

1. Repat the proof of Theorem 2. Consider the sum S defined by

$$S = \sum_{x=N+1}^{2N} e^{2\pi i m f(x)}, \quad p^{9/10} < N < P.$$

Let $a = [N^{5/11}]$ and $r = c_1(\log N)^{\gamma-1}$. Expand the function $mf(n + xy)$ in a Taylor series in powers of xy, and estimate, as in Theorem 2, the double sum W

$$W = \sum_{x=1}^{a} \sum_{y=1}^{a} 2\pi i F(xy), \quad F(xy) = \sum_{s=1}^{r} \alpha_s x^s y^s.$$

(It is nontrivial to estimate the sum of the minima for those s which lie in the interval $e_2(\log N)^{\gamma-1} < s < c_3(\log N)^{\gamma-1}$, where c_1, c_2, c_3 are selected by using the hypotheses of the exercise.)

2. This follows from exercise 1 and Lemma B of Chapter I.

3. This follows from Theorem 2 and from Chapter IV, Theorem 6, by applying partial summation.

4. Apply exercise 1 of Chapter V to the function $f(s) = \zeta^k(s)$. Use exercise 7 of Chapter III to estimate $A(\xi)$. Let $\alpha = 1 - (2ck)^{-2/3}$, where $c > 0$ is the constant in exercise 3, and let $T = X^{1-\alpha}$. Consider the contour integral around the rectangle with vertices $b + iT$, $\alpha + iT$, $\alpha - iT$, where $b = 1 + (\log X)^{-1}$. Since $A(\xi)$ is a nondecreasing function of ξ, then for $h = X^{(\alpha+1)/2}$

$$\frac{1}{h}\int_{X-h}^{X} A(\xi)d\xi \le A(X) \le \frac{1}{h}\int_{X}^{X+h} A(\xi)d\xi.$$

(Cf. H.-E. Richert, Einführung in die Theorie der starken Rieszschen Summierbarkeit von Dirichletreihen, Nachr. Akad. Wiss. Göttingen Math.-Phys. Kl. II (1960), 17–75; A.A. Karatsuba, Estimates of trigonometric sums by the method of I.M. Vinogradov, and their applications, Trudy Mat. Inst. Steklov 112 (1971), 241–255; also see the article referred to in exercise 1 of Chapter V.)

5. For each $j = 1, \ldots, N$, construct the function $\varphi_j(x)$ such that $\varphi_j(x_j) = 0$ for $|x_j - \beta_j| \ge \varepsilon$ and $\varphi_j(\beta_j) = 0$ (cf. Lemma A of Chapter I). Expand the function

$$f(x_1, \ldots, x_N) = \varphi_1(x_1) \ldots \varphi_N(x_N)$$

in a Fourier series and prove that

$$\int_0^T f(\alpha_1 t, \ldots, \alpha_N t)dt = T\int_0^1 \ldots \int_0^1 f(x_1, \ldots, x_N)dx_1 \ldots dx_N + o(T).$$

6. Consider the following two functions of a complex variable:

$$\Phi(X: s_0 + z, \vec{\theta}) \quad \text{and} \quad \Phi(X; s_0 + z).$$

Apply the principle of the argument and exercise 5.

7. Use the Euler product for $\zeta(s)$.

8. If we use relation (4) from section 1 of Chapter 5, we can carry out the computation.

9. The function $F(\theta_{p_1}, \theta_{p_2}, \ldots, \theta_{p_k})$ can be represented in the form

$$F = \sum_{n=1}^{k} \frac{1}{p_n} e^{2\pi i\theta_{p_n}} + O(1) = L(\theta_{p_1}, \ldots, \theta_{ph}) + O(1).$$

Moreover, $O(1)$ is a continuous function of the arguments $\theta_{p_1}, \ldots, \theta_{p_k}$. For

$k > 10$, the set of values of $L(\theta_{p_1}, \ldots, \theta_{p_k})$ is a circle of radius

$$\sum_{n=1}^{k} \frac{1}{p_n}.$$

Because of the divergence of the series Σp^{-1} (cf. exercise 9 of Chapter V), the assertion of the exercise is a consequence of the continuity of the function F.

10. Apply the previous exercise, and change a finite number of the $\lambda(p)$ so that conditions a), b), c), and d) are satisfied. Condition e) can be proved by complex integration (cf. Chapter V).

11. See the hint for the solution of exercise 9.

12. Apply parts e) α) and 3) e) of exercise 10, and also Lagrange's theorem on finite differences.

13. For $X > X_1$, consider

$$F_2(\vec{\theta}) = \sum_{n \leq X} \frac{\lambda'(n)}{n^\sigma} + \sum_{\frac{1}{2}X < p \leq X} \left(\frac{1}{p^\sigma} + \frac{1}{p^\sigma} e^{2\pi i \theta_p} \right),$$

and apply exercises 11 and 12.

14. Use exercises 13 and 6.

(In connection with exercises 5–14, see S.M. Voronin, The zeros of the partial sums of the Dirichlet series of the Riemann-zeta function, Dokl. Akad. Nauk SSSR 216 (1974), 964–967.

Chapter VII

1. To prove B, apply Theorem 1 of Chapter V and use A. To prove A, use the result of exercise 2 in Chapter IV, and apply B.

2. Necessity: a) and b) follow from exercise 1; c) follows from d); to prove d), first obtain the estimate

$$\left| \sum_{n \leq X} \tau_k(n) n^{it} \right| \ll \sqrt{X} |t|^\varepsilon, \quad 1 \ll X \leq |t|^k$$

(cf. exercise 1, B); e) follows from Theorem 1 of Chapter V.

Sufficiently: Suppose that a) is satisfied, and suppose that for some $\varepsilon_0 > 0$ there exists a sequence of integers $T_j \to +\infty$ such that

$$\left| \zeta \left(\frac{1}{2} + iT_j \right) \right| \geq T_j^{\varepsilon_0}$$

It follows from Theorem 6 of Chapter IV that for $|t - T_j| \leq 1/2$

$$\zeta \left(\frac{1}{2} + it \right) = \zeta \left(\frac{1}{2} + iT_j \right) + O \left(|t - T_j| T_j^{1/4} \ln T_j \right) + O(1),$$

i.e.

$$\left| \zeta \left(\frac{1}{2} + it \right) \right| \gg T_j^{\varepsilon_0} \quad \text{for } |t - T_j| \ll T_j^{-1/4}.$$

Therefore,

$$\int_1^{2T_j} \left| \zeta\left(\frac{1}{2} + it\right) \right|^{2k} dt \gg T_j^{2k\varepsilon_0 - 1/4}.$$

We obtain a contradiction for $k > 1/\varepsilon_0$.

The sufficiency of b) and c) is proved similarly. Statement d) is trivial. To prove e), for $\operatorname{Re} s > 1$,

$$\zeta^k(s) = \sum_{n=1}^{\infty} \frac{\tau_k(n)}{n^s} = s \int_1^{\infty} T_k(X) X^{-s-1} dX$$

$$= s \int_1^{\infty} X P_{k-1}(\ln X) X^{-s-1} dX + s \int_1^{\infty} R_k(X) X^{-s-1} dX = f(s) + g(s).$$

The function $f(s)$ is regular for $\operatorname{Re} s > 0$, $s \ne 1$, and the function $g(s)$ is regular for $\operatorname{Re} s > 1/2$. Estimating $|f(s)|$ and $|g(s)|$ for $\operatorname{Re} s \ge s_0 > 1/2$, $|t| \ge 2$, we find

$$|\zeta^k(s)| \ll |t|; \quad |\zeta(\sigma + it)| \ll |t|^{1/k}.$$

for any $k \ge 2$. This implies b). (Cf. Titchmarsh [9].)

3. Apply exercise 2, d) and repeat the proof of Theorem 1.

4. a) Let $N > H > 0$ and J be the number of solutions of the inequality

$$N - p < p' \le N - p + H, \quad p \le N/2$$

i.e.

$$J = \sum_{p \le N/2} (\psi(N - p + H) - \psi(N - p))$$

$$= H\pi\left(\frac{N}{2}\right) - \sum_{p \le N/2} \sum_{|\operatorname{Im}\rho| \le T} \frac{(N - p + H)^{\rho} - (N - p)^{\rho}}{\rho} + O\left(\frac{N^2 \ln N}{T}\right).$$

The absolute value of the second sum does not exceed ω:

$$\omega = \sum_{p \le N/2} \int_{N-p}^{N-p+H} \left| \sum_{|\operatorname{Im}\rho| \le T} x^{\rho-1} \right| dx \ll H N^{-1} \int_{N/2}^{N+H} \left| \sum_{|\operatorname{Im}\rho| \le T} x^{\rho} \right| dx.$$

The last integral can be estimated by using Cauchy's inequality and Theorem 1.

b) This follows from exercise 3, similar to a).

5. Repeat the proof of Theorem 2. (Cf. Yu. V. Linnik, Some conditional theorems concerning the binary Goldbach problem, Izv. Akad. Nauk SSSR Ser. Mat. 16 (1952), 503–520.)

Chapter VIII

1. Consider the system of n linear equations in the $n + 1$ unknown numbers $q_0, q_1, \ldots, q_n$:

$$\begin{cases} \alpha_1 q_n + \alpha_2 q_{n-1} + \ldots + \alpha_{n+1} q_0 = 0, \\ \alpha_2 q_n + \alpha_3 q_{n-1} + \ldots + \alpha_{n+2} q_0 = 0, \\ \cdots\cdots\cdots\cdots\cdots\cdots\cdots\cdots \\ \alpha_n q_n + \alpha_{n+1} q_{n-1} + \ldots + \alpha_{2n} q_0 = 0. \end{cases}$$

2. 1) Let $k_1 + \ldots + k_r \geq n + 1$. Without loss of generality, we can assume that $k_1 + \ldots + k_r = n + 1$. Multiplying out, we choose successively b, c, $d, \ldots, l$, m that satisfy the relation

$$a_n(x - a_1)^{k_1}(x - a_2)^{k_2} \ldots (x - a_r)^{k_r - 1} + b(x - a_1)^{k_1}(x - a_2)^{k_2} \ldots$$

$$(x - a_r)^{k_r - 2} + \ldots + c(x - a_1)^{k_1} + d(x - a_1)^{k_1 - 1} + \ldots$$

$$+ l(x - a_1) + m \equiv F(x) \;(\text{mod } p).$$

It follows from the definition that if

$$(x - a)G(x) \equiv 0 \;(\text{mod } p),$$

then $G(x) \equiv 0 \;(\text{mod } p)$. Next, setting $x = a_1$ in the first expression, we find that $m \equiv 0 \;(\text{mod } p)$. Cancelling the term $(x - a_1)$, and repeating the same argument used above, we find that $l \equiv 0, \ldots, d \equiv 0$, $c \equiv 0$, $b \equiv 0$ modulo p. Finally, it follows from

$$a_n(x - a_1)^{k_1}(x - a_2)^{k_2} \ldots (x - a_r)^{k_r - 1} \equiv (x - a_r)^{k_r} G(x)(\text{mod } p).$$

that $a_n \equiv 0 \;(\text{mod } p)$, which contradicts the hypothesis of the exercise.

2. 2) $F(x) \equiv F(x) - F(a) = (x - a)G(x) \;(\text{mod } p)$.

2. 3) Expanding $F(x)$ in a Taylor series in powers of $(x - a)$, we obtain

$$F(x) \equiv F(x) - F(a) - \frac{F'(a)}{1!}(x - a) - \frac{f''(a)}{2!}(x - a)^2 - \ldots$$

$$- \frac{F^{(k-1)}(a)}{(k-1)!}(x - a)^{k-1} \equiv (x - a)^k G(x)(\text{mod } p).$$

3. (A.G. Postnikov) Use exercise 2. 3) for $k = 1$. If J_1, J_2, J_0 denote the number of solutions, respectively, of the congruences

$$(F(x))^{(p-1)/2} + 1 \equiv 0 \;(\text{mod } p),$$

$$- \{(f(x))^{(p-1)/2}\} + 1 \equiv 0 \;(\text{mod } p),$$

$$f(x) \equiv 0 \;(\text{mod } p),$$

then $N_p = 2J_2 + J_0$, and, by exercise 2. 1),

$$2J_1 + J_0 \leq \deg g(x) = (3/2)(p + 1),$$

$$2J_2 + J_0 \leq \deg g(x).$$

Moreover, $J_1 + J_2 + J_0 = p$. (Cf. L.P. Postnikova, Trigonometric sums and the theory of congruences modulo a prime, in student manual, Moskov. Gos. Ped. Inst., 1973; A. Thue, Über Annäherungswerte algebraischer Zahlen, J. reine angew. Math. 135 (1909), 284–305.)

4. We have

$$F(x^p) = F(x + H) = a_0 + a_1 H + a_2 H^2 + \ldots + a_r H^r,$$

where

$$r = \frac{n(p-1)}{2}, \quad a_0 = F(x), \quad a_v = \frac{1}{v!}F^{(v)}(x) = (f(x))^{\frac{p-1}{2}-v}B_v(x),$$

$$B_v = B_v(x) = C_r^v a^v x^{(n-1)v} + \dots, \quad v = 1, 2, \dots, (p-1)/2.$$

Applying exercise 1, we find a polynomial $h(x)$ of the form

$$h = h(x) = b_0 + b_1 H + \dots + b_k H_k$$

and such that

$$F(x)h(x) = c_0 + c_1 H + \dots + c_k H^k + c_{2k+1} H^{2k+1} + \dots$$

The coefficients $c_0, c_1, \dots, c_k$ are defined by the system of linear equations

$$\begin{cases} c_0 = a_0 b_0, \\ c_1 = a_0 b_1 + a_1 b_0, \\ \dots\dots\dots\dots\dots\dots\dots\dots\dots\dots \\ c_k = a_0 b_k + a_1 b_{k-1} + \dots + a_k b_0, \\ 0 = a_1 b_k + a_2 b_{k-1} + \dots + a_{k+1} b_0, \\ \dots\dots\dots\dots\dots\dots\dots\dots\dots\dots \\ 0 = a_k b_k + a_{k+1} b_{k-1} + \dots + a_{2k} b_0. \end{cases}$$

The last n equations in this system can be described as follows, with $f(x) = f$:

$$\begin{cases} f^k B_1 b_k + f^{k-1} B_2 b_{k-1} + \dots + B_{k+1} b_0 = 0, \\ f^k B_2 b_k + f^{k-1} B_3 b_{k-1} + \dots + B_{k+2} b_0 = 0, \\ \dots\dots\dots\dots\dots\dots\dots\dots\dots\dots\dots\dots \\ f^k B_k b_k + f^{k-1} B_{k+1} b_{k-1} + \dots + B_{2k} b_0 = 0. \end{cases}$$

From this system we can determine $b_0, b_1, \dots, b_k$:

$$b_{k-s+1} = (-1)^s f^{s-1} \begin{vmatrix} B_1 & \dots & B_{s-1} & B_{s+1} & \dots & B_{k+1} \\ B_2 & \dots & B_s & B_{s+2} & \dots & B_{k+2} \\ \multicolumn{6}{c}{\dots\dots\dots\dots\dots\dots\dots\dots\dots} \\ B_k & \dots & B_{s+k-2} & B_{s+k} & \dots & B_{2k} \end{vmatrix}.$$

The leading term of the polynomial b_{k-s+1} has the form

$$(-1)^s a^{k(k-1)} x^{k(k+1)(n-1)+s-1} \begin{vmatrix} C_r^1 & \dots & C_r^{s-1} & C_r^{s+1} & \dots & C_r^{k+1} \\ \multicolumn{6}{c}{\dots\dots\dots\dots\dots\dots\dots\dots\dots} \\ C_r^k & \dots & C_r^{k+s-2} & C_r^{k+s} & \dots & C_r^{2k} \end{vmatrix}.$$

The degree of the polynomial $a_s(x)$ is $(n-s)(p-1)/2$, and the coefficient of the highest power of $a_s(x)$ is $a^{(p-1)/2}$. The degree of c_k does not exceed the degree of $a_s b_{k-s}$ for $0 \le s \le k$, and for any s, $0 \le s \le k$, the degree of $a_s b_{k-s}$ is equal to

$$k(k+1)(n-1) + \frac{p-1}{2}n.$$

The coefficient of this degree in the polynomial c_k is equal to x:

$$\chi = \sum_{s=1}^{k+1} (-1)^{s-1} a^{k(k+1) + \frac{\ell-1}{2}} \begin{vmatrix} C_r^1 \dots C_r^{s-1} & C_r^{s+1} \dots C_r^{k+1} \\ \dots \dots \dots \dots \dots \dots \dots \\ C_r^k \dots C_r^{k+s-2} & C_r^{k+s} \dots C_r^{2k} \end{vmatrix}$$

$$= a^{k(k+1) + \frac{\ell-1}{2}} \begin{vmatrix} 1 & 1 & \cdots 1 \\ C_r^1 & C_r^2 & \cdots C_r^{k+1} \\ \dots \dots & \dots \dots \\ C_r^k & C_r^{k+1} \cdots C_r^{2k} \end{vmatrix}.$$

Since n is an odd integer, and $k \leq (p-1)/4$, then $x \not\equiv 0 \pmod p$. Thus, the polynomial $g(x)$

$$g(x) = c_0 + c_1 H + \dots + c_k H^k,$$

has degree m modulo p, where

$$m = kp + k(k+1)(n-1) + \frac{p-1}{2} n,$$

and each root of $F(x)$ modulo p is a root of multiplicity $2k+1$ of the polynomial $g(x)$.

5. We shall assume that $n \geq 3$ and $p > 4n^2$. Let J_v be the number of solutions of the congruence

$$F_v(x) = f^{\frac{p-1}{2}}(x) + (-1)^v \equiv 0 \pmod p, \quad v = 1, 2.$$

Then

$$S = \sum_{x=1}^{p} \left(\frac{f(x)}{p} \right) = J_1 - J_2.$$

In exercise 4, choose $k = [(\sqrt{2p} - 1)/2] + 1$, and apply exercise 2. We get

$$(2k+1)J_v \leq m = kp + k(k+1)(n-1) + \frac{p-1}{2} n,$$

$$J_v \leq \frac{1}{2} p + (n-1) \left(\sqrt{\frac{p}{2}} + \frac{1}{2} \right).$$

Moreover,

$$J_1 + J_2 \geq p - n.$$

Consequently,

$$J_2 \geq p - n - J_1 \geq \frac{1}{2} p - (n-1) \left(\sqrt{\frac{p}{2}} + \frac{1}{2} \right) - n;$$

$$S| \leq |J_1 - J_2| \leq 2(n-1) \left(\sqrt{\frac{p}{2}} + \frac{1}{2} \right) + n < 2n\sqrt{p}.$$

(In connection with exercises 4–5, cf. S.A. Stepanov, The number of points of a hyperelliptic curve over a finite prime field, Izv. Akad. Nauk SSSR Ser. Mat. 33 (1969), 1171–1181; N.M. Korobov, An estimate for a sum of Legendre symbols, Dokl. Akad. Nauk SSSR 96 (1971), 764–767.)

6. For $X < p$, we have

$$V(X) = \frac{1}{2} \sum_{n \leq X} \left(\left(\frac{n}{p} \right) + 1 \right) = \frac{1}{2} X + \frac{1}{2} \sum_{n \leq X} \left(\frac{n}{p} \right) + \theta_1,$$

$$N(X) = \frac{1}{2} X - \frac{1}{2} \sum_{n \leq X} \left(\frac{n}{p} \right) + \theta_2.$$

The exercise now follows from Lemma 5 of Chapter VIII.

7. The non-residues among the integers $1, 2, \ldots, Y$ will be the numbers that are divisible by a prime greater than n, i.e.

$$N(Y) \leq \sum_{n < p \leq Y} \frac{Y}{p} = Y \left(\ln \frac{\ln Y}{\ln n} + O \left(\frac{1}{\ln n} \right) \right).$$

On the other hand,

$$N(Y) = (1/2) Y + o(Y),$$

i.e.

$$\frac{1}{2} + o(1) \leq \ln \frac{\ln Y}{\ln n} + O \left(\frac{1}{\ln n} \right).$$

(Cf. I.M. Vinogradov, Elements of Number Theory (Russian), Nauka, Moscow, 1981, p. 113.)

8. We have

$$\sum_{\lambda=1}^{p} \left(\sum_{m=1}^{z} \left(\frac{\lambda + m}{p} \right) \right)^{2k} = \sum_{m_1, \ldots, m_{2k} = 1}^{z} \sum_{\lambda=1}^{p} \left(\frac{(\lambda + m_1) \cdots (\lambda + m_{2k})}{p} \right).$$

The $2k$-tuples $(m_1, \ldots, m_{2k})$ can be divided into two classes A and B: The class A consists of all $2k$-tuples that have at least $k + 1$ different values of m_j, and the class B consists of all of the remaining $2k$-tuples. The number of $2k$-tuples in class B does not exceed $k! Z^k$. If the $2k$-tuple $(m_1, \ldots, m_{2k})$ belongs to class A, and

$$(\lambda + m_1) \ldots (\lambda + m_{2k}) \neq (\lambda + m_1')^{\alpha_1} \ldots (\lambda + m_r')^{\alpha_r},$$

where $m_1', \ldots, m_r'$ are pairwise distinct integers, then one of the α_j must equal 1. Now apply the estimate in exercise 5 to the sum

$$\sum_{\lambda=1}^{p} \left(\frac{(\lambda + m_1')^{\alpha_1} \ldots (\lambda + m_r')}{p} \right).$$

9. Let $k = [(2\varepsilon)^{-1}] + 1$, raise $|W|$ to the power $2k$, and apply Hölder's inequality and exercise 8. (Cf. A.A. Karatsuba, Sums of characters with primes

that belong to an arithmetic progression, Izv. Akad. Nauk SSSR Ser. Mat. 35 (1971), 469–484, Lemma 4.)

10. Let $Y = [p^{0.25(1-\varepsilon)}]$ and $Z = [p^{0.25\varepsilon}]$. Then (cf. the proof of Theorem 2, Chapter VI)

$$S = (YZ)^{-1} \sum_{m \leq X} \sum_{y \leq Y} \sum_{z \leq Z} \left(\frac{m + yz}{p} \right) + O(Xp^{-\varepsilon});$$

$$W \leq (YZ)^{-1} \sum_{m \leq X} \sum_{y \leq Y} \left| \sum_{z \leq Z} \left(\frac{my' + z}{p} \right) \right|,$$

where $yy' \equiv 1 \pmod{p}$. Furthermore, for any $k \geq 1$,

$$W^k \leq (YZ)^{-k}(XY)^{k-1} \sum_{m \leq X} \sum_{y \leq Y} \left| \sum_{z \leq Z} \left(\frac{my' + z}{p} \right) \right|^k$$

$$= X^{k-1} Y^{-1} Z^{-k} \sum_{\lambda=1}^{p-1} \tau'(\lambda) \left| \sum_{z \leq Z} \left(\frac{\lambda + z}{p} \right) \right|^k,$$

where $\tau'(\lambda)$ is the number of solutions of the congruence

$$my' \equiv \lambda \pmod{p}, \quad m \leq X, \quad y \leq Y.$$

We apply Cauchy's inequality:

$$W^{2k} \leq X^{2k-2} Y^{-2k} Z^{-2} \sum_{\lambda=1}^{p-1} (\tau'(\lambda))^2 \sum_{\lambda=1}^{p-1} \left| \sum_{z \leq Z} \left(\frac{\lambda + z}{p} \right) \right|^{2k}.$$

The first sum in the last inequality is equal to the number of solutions of the congruence $my_1 \equiv m_1 y \pmod{p}$, where $m, m_1 \leq X$ and $y, y_1 \leq Y$ and this, in turn, does not exceed

$$\sum_{\lambda \leq XY} \tau^2(\lambda) = O(XY \ln^3 p).$$

Apply exercise 8 to the second sum. Choosing $k = k(\varepsilon)$ completes the proof. (Cf. D.A. Burgess, The distribution of quadratic residues and non-reidues, Mathematika 4 (1957), 106–112; A.A. Karatsuba, Sums of characters and primitive roots in finite fields, Dokl. Akad. Nauk SSSR 180 (1968), 1287–1289.)

11. This follows from exercises 7 and 12.

12. This follows from the Lemma in Chapter VII.

13. (S. Uchiyama) Let $0 < \varepsilon < 0{,}01$, $Y = \ln^{2+\varepsilon} X$, $k = \dfrac{(2 - \varepsilon)\ln X}{2\ln \ln X}$, $X \geq X_0(\varepsilon)$. Consider the sum W_m,

$$W_m = \sum_{p \leq X} \left(\sum_{q \leq Y} \left(\frac{q}{p} \right) \right)^{2m}.$$

where m is a natural number and q and p are prime numbers. We have

$$W_m = \sum_{p \leq X} \left| \sum_{n \leq Y^m} \tau'_m(n) \left(\frac{n}{p} \right) \right|^2,$$

where $\tau'_m(n)$ is the number of solutions of the equation $q_1 \ldots q_m = n$ in prime numbers $q_j \leq Y, j = 1, \ldots, m$. Applying formula (5) and exercise (12), we obtain the inequality

$$W_m \leq \sum_{p \leq X} \sum_{a=1}^{p-1} \left| \sum_{n \leq Y^m} \tau'_m(n)e^{2\pi i^{an/p}} \right|^2$$

$$\leq c(Y^m + X^2) \sum_{n \leq Y^m} (\tau'_m(n))^2 \leq cm^m Y^m(Y^m + X^2).$$

Now let M have the value p, $p \leq X$. Then

$$\left| \sum_{q \leq Y} \left(\frac{q}{p} \right) \right|^{2k} = (\pi(Y))^{2k}.$$

The preceding inequality implies that

$$M(\pi(Y))^{2k} \leq ck^k Y^k(Y^k + X^2); \quad M \leq c_1 \frac{X}{\ln^{1+\delta} X}, \quad \delta = \delta(\varepsilon) > 0.$$

Consequently, for $\pi(X) + O\left(\dfrac{X}{\ln^{1+\delta} X} \right)$ values of $p \leq X$ we have the inequality

$$\left| \sum_{q \leq Y} \left(\frac{q}{p} \right) \right| < \pi(Y),$$

i.e. among the numbers $q \leq Y$ these are quadratic residues and non-residues modulo p.

Chapter IX

1. a) We have

$$\sum_{Z < n \leq N} \frac{\chi(n)}{n^s} = \sum_{Z < n \leq N} \frac{1}{n^\sigma} \chi(n)n^{-it} = \sigma \int_Z^N C(u)u^{-1-\sigma}du + C(N)N^{-\sigma},$$

where

$$C(u) = \sum_{Z < n \leq u} \chi(n)n^{-it} = \sum_{l=0}^{k-1} \chi(l) \sum_{(Z-l)k^{-1} < m \leq (u-l)k^{-1}} e^{-it \log(mk+l)}.$$

Applying the Corrollary to Lemma 1 of Chapter I to the sum over m, we obtain

$$C(u) = \sum_{l=0}^{k-1} \chi(l)\left(\int_2^u x^{-it}dx + O(1) \right) = O(k).$$

This together with the first formula as $N \to +\infty$ completes the proof.

b) Replace the function $\zeta(s)$ by $L(s, \chi)$, we repeat the proof of exercise 1 in Chapter IV.

2. Since χ is a primitive character modulo k, then by Lemma 3 of Chapter VIII

$$\chi(n) = \frac{1}{\tau(\bar{\chi})} \sum_{b=1}^k \bar{\chi}(b)e^{2\pi i^{bn/k}}, \quad |\tau(\bar{\chi})| = \sqrt{k}.$$

Extending in the appropriate place the summation over all characters modulo k and using the multiplicativity of the characters, we obtain

$$\sum_{\chi \bmod k}{}' \left| \sum_{n=M+1}^{M+N} a_n \chi(n) \right|^2$$

$$\leq \frac{1}{k} \sum_{\chi \bmod k} \sum_{n=M+1}^{M+N} \sum_{m=M+1}^{M+N} a_n \bar{a}_m \sum_{\substack{b=1 \\ (b,k)=1}}^{k} \bar{\chi}(b) \sum_{\substack{c=1 \\ (c,k)=1}}^{k} \chi(c) e^{2\pi i \frac{bn-cm}{k}}$$

$$= \frac{\varphi(k)}{k} \sum_{\substack{b=1 \\ (b,k)=1}}^{k} \sum_{n=M+1}^{M+N} \sum_{m=M+1}^{M+N} a_n \bar{a}_m e^{2\pi i \frac{b(n-m)}{k}}$$

$$= \frac{\varphi(k)}{k} \sum_{\substack{b=1 \\ (b,k)=1}}^{k} \left| \sum_{n=M+1}^{M+N} a_n e^{2\pi i bn/k} \right|^2.$$

Next, we apply exercise 12 of Chapter VIII to complete the proof.

3. We use

$$\sum_{n=M+1}^{M+N} a_n \chi(n) n^{-iA} n^{-s_\chi + iA}$$

$$= (s_\chi - iA) \int_{M+1}^{M+N} C(u) u^{-s_\chi + iA - 1} du + C(M+N)(M+N)^{-s_\chi + iA},$$

where

$$C(u) = \sum_{M < n \leq u} a_n \chi(n) n^{-iA};$$

$$\left| \sum_{n=M+1}^{M+N} a_n \chi(n) n^{-s_\chi} \right|^2 \leq 20 \left(\int_{M+1}^{M+N} |C(u)| u^{-1} du \right)^2$$

$$+ 20|C(M+N)|^2 \leq 20L \int_{M+1}^{M+N} u^{-1} |C(u)|^2 du + 20|C(M+N)|^2.$$

We complete the proof by summing over both sides of the inequality by χ and k, and applying exercise 2.

4. Multiply equation a) of exercise 1 by

$$M_\chi(s, \chi) = \sum_{n \leq X} \mu(n) \chi(n) n^{-s}.$$

Using the properties of the Möbius function, we obtain for $s = \rho$

$$0 = 1 + \sum_{X < n \leq XY} \frac{a(n)\chi(n)}{n^\rho} + \left(\sum_{Y < n \leq Z} \frac{\chi(n)}{n^\rho} \right) M_X(\rho, \chi) + O(kZ^{-\sigma}|M_X(\rho, \chi)|),$$

where

$$a(n) = \sum_{\substack{d \backslash n \\ nY \leq d \leq \min(X, nX^{-1})}} \mu(d), \quad |a(n)| \leq \tau(n).$$

The exercise follows trivially from this.

5. Let $s = \rho$ be a zero of $L(s, \chi)$, and let $\mathrm{Re}\, s \geq \alpha$ and $|\mathrm{Im}\, s| \leq T$. In exercise 4, let

$$X = B^{\frac{1}{2(3-2\alpha)}}, \quad Y = B^{\frac{3}{2(3-2\alpha)}}, \quad B = Q^3 + QT,$$

$$Z = \begin{cases} B, & \text{if } \frac{1}{2} \leq \alpha < \frac{3}{4}; \\ Y, & \text{if } \frac{3}{4} \leq \alpha < 1. \end{cases}$$

All of the zeros we are considering for all of the L-functions with primitive character χ modulo $k \leq Q$ can be divided into four classes, corresponding to the four inequalities in exercise 4 (the classes can intersect). Summing each of the inequalities over the zeros in its class, and then adding the four resulting inequalities, leads to the following relation:

$$\sum_{k \leq Q} \sideset{}{'}\sum_{\chi \bmod k} N(\alpha, T, \chi) \leq c_2 \sum_{k \leq Q} \sideset{}{'}\sum_{\chi \bmod k} \sum_{\rho_x} (\kappa_1 + \kappa_2 + \kappa_3 + \kappa_4).$$

Each sum on the right can be estimated like the sum in exercise 3: the zeros ρ_x are divided into classes defined by

$$A \leq \mathrm{Im}\, \rho_x < A + 1,$$

$A = -T, -T+1, \ldots, T-2, T-1$. Each class has $\ll \ln(Q+T)$ zeros, and the number of classes is $\ll T$. To estimate the third sum, we use Hölder's inequality again:

$$\sum_{k \leq Q} \sideset{}{'}\sum_{\chi \bmod k} \left| \sum_{n \leq X} \mu(n)\chi(n)n^{-\rho} \right|^{4/3} \left| \sum_{Y < n \leq Z} \chi(n)n^{-\rho} \right|^{4/3}$$

$$\leq \left(\sum_{k \leq Q} \sideset{}{'}\sum_{\chi \bmod k} \left| \sum_{n \leq X} \mu(n)\chi(n)n^{-\rho} \right|^2 \right)^{2/3} \left(\sum_{k \leq Q} \sideset{}{'}\sum_{\chi \bmod k} \left| \sum_{Y < n \leq Z} \chi(n)n^{-\rho} \right|^4 \right)^{1/3}.$$

These sums can all be estimated in the same way. For example, we shall estimate the last sum:

$$\Sigma = \sum_{k \leq Q} \sideset{}{'}\sum_{\chi \bmod k} \left| \sum_{Y < n \leq Z} \chi(n)n^{-\rho} \right|^4 = \sum_{k \leq Q} \sideset{}{'}\sum_{\chi \bmod k} \left| \sum_{Y^2 < n \leq Z^2} \tau'(n)\chi(n)n^{-\rho} \right|^2,$$

where $\tau'(n) \leq \tau(n)$, $\rho = \rho_x$, $A \leq \mathrm{Im}\, \rho_x < A + 1$. For this we consider the sum W,

$$W = \sum_{k \leq Q} \sideset{}{'}\sum_{\chi \bmod k} \left| \sum_{N < n \leq N_1 \leq 2N} \tau'(n)n^{-\alpha}\chi(n)n^{\alpha-\rho} \right|^2.$$

This sum corresponds to the sum of exercise 3, c)

$$a_n = \tau'(n)n^{-\alpha}.$$

Therefore,

$$W \leq c(Q^2 + N) \sum_{n=N+1}^{2N} n^{-2\alpha}\tau^2(n) \leq c_1(Q^2 + N)N^{-2\alpha+1}\ln^3(Q+T).$$

Similarly,

$$\Sigma \leq c_2(Q^2 Y^{2(1-2\alpha)} + Z^{4(1-\alpha)})\ln^5(Q+T).$$

The assertion of the exercise follows from some calculations that are not too complicated.

6. a) From §1

$$\psi(x; k, l) - \frac{\psi(x)}{\varphi(k)} = \frac{1}{\varphi(k)} \sum_{\chi \neq \chi_0} \psi(x, \chi)\bar{\chi}(l) + O(\ln^2 x);$$

$$\psi(x, \chi) = \psi(x, \chi_1) + O(\ln^2 x),$$

where χ_1 is the primitive character modulo k_1 that induces the character χ, and $k_1 | k$. Then

$$\sigma = \sum_{k \leq \sqrt{X}(\ln X)^{-B}} \max_{(l, k) = 1} \left| \psi(X; k, l) - \frac{X}{\varphi(k)} \right|$$

$$\leq \sum_{k \leq \sqrt{X}(\ln X)^{-B}} \frac{1}{\varphi(k)} \sum_{\chi \neq \chi_0} |\psi(X, \chi)| + O(\sqrt{X}(\ln X)^{-B+2})$$

$$\leq \sum_{k_1 \leq \sqrt{X}(\ln X)^{-B}} \sum_{k_1 r \leq \sqrt{X}(\ln X)^{-B}} \frac{1}{\varphi(k_1 r)} \sum_{\chi_1 \bmod k_1} (|\psi(X, \chi_1)|$$

$$+ O(\ln^2 X)) + O(\sqrt{X}(\ln X)^{-B+2}).$$

Since

$$\varphi(k_1 r) \geq \varphi(k_1)\varphi(r), \quad \varphi(r) > cr(\ln \ln X)^{-1},$$

then

$$\sigma \ll (\ln^2 X) \sum_{k_1 \leq \sqrt{X}(\ln X)^{-B}} \frac{1}{\varphi(k_1)} \sum_{\chi_1 \bmod k_1} |\psi(X, \chi_1)| + \sqrt{X}(\ln X)^{-B+2}.$$

If $k_1 \leq (\ln X)^N$, where $N > 0$ is any fixed integer, then the required estimate follows from Theorem 6. Let

$$(\ln X)^N < k_1 \leq \sqrt{X}(\ln X)^{-B}.$$

Consider $\sigma(K)$,

$$\sigma(K) = \sum_{K < k_1 \leq 2K} \frac{1}{\varphi(k_1)} \sum_{\chi_1 \bmod k_1} |\psi(X, \chi_1)|,$$

By Theorem 1 for $T = K(\ln X)^{A+10}$

$$|\psi(X, \chi_1)| \ll \sum_{|\mathrm{Im}\,\rho| \leq T} \frac{|X^\rho|}{|\rho|} + \frac{X}{K(\ln X)^{A+8}}.$$

Now we estimate the sum:

$$\sigma_1 = \sum_{K < k_1 \leq 2K} \sum_{\chi_1 \bmod k_1} \sum_{T_1 < \mathrm{Im}\,\rho \leq 2T_1} X^\sigma,$$

where $\rho = \sigma + it$. We have

$$X^\sigma = X^{1/2} + (\ln X) \int\limits_{0,5}^{\sigma} X^\alpha d\alpha,$$

$$\sum_{|\operatorname{Im}\rho| \le 2T} X^\sigma = X^{1/2} N\left(\frac{1}{2}, 2T_1, \chi\right) + (\ln X) \int\limits_{0,5}^{1} X^\alpha N(\alpha, 2T_1, \chi) d\alpha.$$

Therefore,

$$\sigma_1 \ll (\ln X) \max_\alpha X^\alpha \sum_{K < h_1 \le 2K} \sum_{\chi_1 \bmod k_1} N(\alpha, 2T_1, \chi_1).$$

Substituting the result of exercise 5 into the last estimate, and then collecting all the preceding estimates, we obtain what is required with $B = A + 2$.

b) Repeat the argument of a) and use, where necessary, Corollaries 2 and 3 to Theorem 4 and Corollary 3 to Theorem 6.

(In connection with exercises 2–6, cf. [1], [4], and [7], and also A.I. Vinogradov, The density hypothesis for Dirichlet L-series, Izv. Akad, Nauk SSSR Ser. Mat. 29 (1965), 903–934; E. Bombieri, On the large sieve, Mathematika 12 (1965), 201–225.)

7. Using the argument of Lemma 3 of Chapter 8, prove that

$$\operatorname{ind}(1 + \rho^s u) = (p - 1)p^{s-1}\gamma, \text{ where } (\gamma, p) = 1.$$

Next, consider the function f

$$f(1 + p^s z) = p^s z - \frac{1}{2}(p^s z)^2 + \cdots + (-1)^{m-1}\frac{1}{m}(p^s z)^m,$$

and prove that

$$f(1 + p^s z_1) + f(1 + p^s z_2) = f((1 + p^s z_1)(1 + p^s z_2))(\bmod \rho^n).$$

It follows from this and from the previously deduced relation that

$$\operatorname{ind}(1 + p^s u) \equiv (p - 1)af(1 + p^s u)(\bmod(p - 1)p^{n-1}),$$

where $(a, p) = 1$. (cf. A.G. Postnikov, On the sum of characters with respect to a modulus equal to a power of a prime number, Izv. Akad. Nauk SSSR Ser. Mat. 19 (1955), 11–16.)

8. This can be estimated in the same way that the zeta sum was estimated in Thoerem 2 of Chapter VI: Letting $s = [0.5nr^{-1}]$ and $a = [N^{0.25}]$, we obtain the inequality

$$|S| \le a^{-2} N|W| + 2a^2,$$

where

$$W = \sum_{x=1}^{a} \sum_{y=1}^{a} \chi(1 + up^s xy), \quad (u, p) = 1.$$

Applying exercise 7, we can replace this sum by a trigonometric sum and then repeat word for word the proof of Theorem 2 of Chapter VI. In the appropriate place we make a nontrivial estimate of the sum of the minima with index m from the interval $0.5r < m < 1.5r$. (Cf. S.M. Rozin, On the zeros of Dirichlet L-series, Izv. Akad. Nauk SSSR Ser. Mat. 23 (1959), 503–508; A.A. Karatsuba, Trigonometric sums of a special type and their applications, Izv. Akad. Nauk SSSR Ser. Mat. 28 (1964), 237–248; V.N. Chubarikov, A more precise boundary for the zeros of the Dirichlet L-series with modulus equal to a power of a prime, Vestnik Moskov. Univ. Ser. I Mat. Mekh, 28 (1973), no. 2 46–52.)

9. Let $\chi \neq \chi_0$. For any integer $X \geq 1$, $\mathrm{Re}\, s = \sigma > 0$, if we use partial summation, we find that

$$L(s, \chi) = \sum_{n \leq X} \chi(n) n^{-s} + s \int_X^\infty C(u) u^{-s-1} \, du,$$

where

$$C(u) = \sum_{n \leq u} \chi(n) = O(k).$$

Indeed,

$$L(s, \chi) = \sum_{n \leq k} \chi(n) n^{-s} + O((|t| + 1) k^{1-\sigma})$$

Dividing the sum over n into two pieces, corresponding to $n \leq N$ and $n > N$, and using in the first case the trivial estimate and in the second case the estimate obtained in problem 8 for $N = \exp(\log k)^{2/3}$ and

$$\mathrm{Re}\, s = \sigma \geq 1 - \frac{c}{(\log k)^{2/3}},$$

we find that

$$L(s, \chi) = O((|t| + 1)(\log k)^{2/3}).$$

Next, for

$$|t| \leq \exp((\log \log k)^2)$$

we repeat the argument used in Theorem 2 of Chapter VI with the parameters

$$\sigma_0 = 1 + \frac{4d}{(\log k)^{2/3}(\log k)^2}, \quad \sigma = 1 - \frac{d}{(\log k)^{2/3}(\log \log k)^2},$$

$$M = (|t| + 1)\frac{\log^2 k}{d}, \quad r = \frac{c_1}{(\log k)^{2/3}}.$$

10. From exercise 5 with

$$k = p^n, \quad \frac{1}{2}x^\delta < k < x^\delta, \quad T = \exp(\ln \ln x)^2$$

we have

$$\sum_{\chi \bmod h} N(\alpha, T, \chi) \ll Tk^{8(1-\alpha)} \ln^{10} x.$$

Consequently,

$$\psi(x; k, l) = \frac{\psi(x)}{\varphi(k)} + O(R) + O\left(\frac{x}{T}\ln^3 x\right),$$

where

$$R = \frac{1}{\varphi(k)} \sum_{\chi \bmod k} \sum_{|\operatorname{Im}\rho| \leq T} \frac{x^{\sigma_\chi}}{|\rho_\chi|}.$$

Applying exercise 9 (cf. the proof of Theorem 2 of Chapter VII), we obtain with $\gamma = 1 - c_1(\ln x)^{-2/3}(\ln \ln x)^{-2}$

$$\sum_{\operatorname{Im}\rho_\chi \leq T_1} x^{\sigma_\chi} \ll x^{1/2} N\left(\frac{1}{2}, T_1, \chi\right) + (\ln x) \int_{0,5}^{\gamma} x^\alpha N(\alpha, T_1, \chi)d\alpha,$$

$$\sum_{\chi \bmod k} \sum_{|\operatorname{Im}\rho_\chi| \leq T_1} x^{\sigma_\chi} \ll {}^{1/2}T_1 k^4 \ln^{10} x + (\ln x)^{11} x T_1 \int_{0,5}^{\gamma} \left(\frac{k^8}{x}\right)^{1-\alpha} d\alpha$$

$$= O(T_1 x \exp(-\ln^{0,25} x)).$$

Partitioning the sum in R over ρ_χ into $\ll \ln T$ sums and applying the last estimate, we obtain what is required. (Cf. M.B. Barban, Yu. V. Linnik, and N.G. Chudakov, On prime numbers in an arithmetic progression with a prime-power difference, Acta Arith. 9 (1964), 375–390.)

Chapter X

1. We can assume without loss of generality that $(n, m) = 1$. Repeating the proof of Vinogradov's theorem on the sum of three primes, we obtain the following formula for the number I of solutions of the equation in the exercise:

$$I = \frac{I_0(N)}{\ln^3 N}\sigma + O\left(\frac{N^2}{\ln^{3.4} N}\right),$$

where $I_0(N)$ is the number of solutions in natural numbers x, y, z of the equation

$$nx + my + kz = N,$$

$$\sigma = \sum_{q=1}^{\infty} \gamma(q), \quad \gamma(q) = \frac{\varkappa_n(q)\varkappa_m(q)\varkappa_k(q)\overline{\varkappa_N(q)}}{\varphi^3(q)},$$

$$\varkappa_r(q) = \sum_{\substack{a=1 \\ (a,q)=1}}^{q} e^{2\pi i \frac{ar}{q}}.$$

Since $\varkappa_r(q)$ is a multiplicative function, then

$$\sigma = \prod_p (1 + \gamma(p) + \gamma(p^2) + \ldots).$$

Moreover, it follows from the condition $(n, m) = 1$ that

$$\gamma(p^s) = 0 \quad \text{for } s \geq 2.$$

Therefore,

$$\sigma = \prod_{p\backslash nmkN} (1 + \gamma(p)) \prod_{p \times nmkN} \left(1 + \frac{1}{(p-1)^3}\right).$$

2. Let a_n be an even integer, $X/2 < a_n \leq X$ for $n = 1, 2, \ldots, N$. Consider the equation $p_1 + p_2 = a_n$, where p_1 and p_2 are primes. Just as in the proof of Theorem 1, we let $L = \ln X$, $\tau = XL^{-A}$, $\chi\tau = 1$, $Q = L^B$, and we define the sets E_1, E_2 and the integrals I_1, I_2. We obtain

$$I = \int_{-\chi}^{1-\chi} S^2(\alpha) T(\alpha) d\alpha = I_1 + I_2,$$

where I is the number of solutions of the equation under consideration, and

$$S(\alpha) = \sum_{p \leq X} e^{2\pi i \alpha p}, \quad T(\alpha) = \sum_{n=1}^{N} e^{-2\pi i \alpha a_n}.$$

By Theorem 2,

$$I_2 \leq \max_{\alpha \in E_2} |S(\alpha)| \int_0^1 |S(\alpha)| \, |T(\alpha)| \, d\alpha$$

$$\ll X(L^{-0.5B} + L^{-0.5A}) L^3 \sqrt{\int_0^1 |S(\alpha)|^2 \, d\alpha \int_0^1 |T(\alpha)|^2 \, d\alpha}$$

$$= O(X^{1.5} N^{0.5} (L^{-0.5B} + L^{-0.5A}) L^3)).$$

Next,

$$I_1 = \sum_{n=1}^{N} J(n),$$

$$J(n) = \sum_{q \leq Q} \sum_{(a, q) = 1} \int_{-(q\tau)^{-1}}^{(q\tau)^{-1}} S^2\left(\frac{a}{q} + z\right) e^{-2\pi i(\frac{a}{q} + z) a_n} dz.$$

Applying the formula from section 2,

$$S\left(\frac{a}{q} + z\right) = \frac{\mu(q)}{\varphi(q)} \sum_{m \leq X} \frac{e^{2\pi i z m}}{\log m} + O(Xe^{-c\sqrt{L}})$$

and applying the same reasoning, we obtain

$$I_1 = \sum_{n=1}^{N} I(a_n)\sigma(a_n) + O(XNL^{-B}) + O(X^{1.5} N^{0.5} L^{-A+B}),$$

where

$$I(a_n) = \sum_{n+m=a_n} \frac{1}{\log n \log m},$$

$$\sigma(a_n) = \prod_{p \times a_n} \left(1 - \frac{1}{(p-1)^2}\right) \prod_{p\backslash a_n} \left(1 + \frac{1}{p-1}\right).$$

From this, for $A = 2D + 12$ and $B = D + 10$, we obtain the result. (Cf. N.G. Chudakov, On Goldbach's problem, Dokl. Akad. Nauk SSSR 17 (1953), 331–334.)

3. Let $a = p_1^{\alpha_1} \ldots p_k^{\alpha_k}$ be the canonical decomposition of a into prime factors, and let $m < k$, m even. Then

$$\sum_{\substack{d \setminus a \\ \Omega(d) \le m}} \mu(d) = \binom{k}{0} - \binom{k}{1} + \ldots + (-1)^m \binom{k}{m}$$

$$= (-1)^m \left(\binom{k}{m+1} - \binom{k}{m+2} + \ldots \right) \ge 0$$

Therefore,

$$S' = \sum_{v=1}^{n} f(z_v) \sum_{d \setminus (z_v, P)} \mu(d) \le \sum_{v=1}^{n} f(z_v) \sum_{\substack{d \setminus (z_v, P) \\ \Omega(d) \le m}} \mu(d) = \sum_{\substack{d \setminus P \\ \Omega(d) \le m}} \mu(d) S_d.$$

4. a) In exercise 3, we take $m = 10[\ln \ln x]$, $x \ge x_0$,

$$P = \prod_{\substack{p \le b \\ (p, k) = 1}} p; \quad \Omega(P) = n.$$

Using exercise 9 of Chapter V, we find that

$$T \le \sum_{\substack{d \setminus P \\ \Omega(d) \le m}} \mu(d) \left(\frac{x}{kd} + \theta_d \right) \le T_1 + T_2 + T_3,$$

where

$$T_1 = \frac{x}{k} \sum_{d \setminus P} \frac{\mu(d)}{d} = \frac{x}{k} \frac{\prod_{p \le b} \left(1 - \frac{1}{p} \right)}{\prod_{p \setminus k} \left(1 - \frac{1}{p} \right)} \ll \frac{x \ln \ln x}{\varphi(k) \ln x};$$

$$T_2 = \frac{x}{k} \sum_{\substack{d \setminus P \\ \Omega(d) > m}} \frac{1}{d} = \frac{x}{k} \sum_{r=m+1}^{n} \sum_{\Omega(d)=r} \frac{1}{d} \le \frac{x}{k} \sum_{r=m+1}^{n} \frac{1}{r!} \left(\frac{1}{2} + \frac{1}{3} + \ldots + \frac{1}{p_n} \right)^r$$

$$\le \frac{x}{k} \sum_{r=m+1}^{n} \left(\frac{\ln \ln b + c_1}{r} e \right)^r \le \frac{x}{k} \sum_{r=m+1}^{n} \left(\frac{1}{3} \right)^r \ll \frac{x}{\varphi(k) \ln x};$$

$$T_3 = \sum_{\substack{d \setminus P \\ \Omega(d) \le m}} 1 = \sum_{r=0}^{m} \binom{n}{r} \le n^m \le e^{m \ln b} \le x^{1/100} \le \frac{x}{\varphi(k) \ln x}.$$

b) Repeat a) with $\ln b = \ln x (e_1 \ln \ln x)^{-1}$, $m = 2[c_2 \ln \ln x]$, where c_1 and c_2 are chosen independent of α as required. (Cf. Prachar [8], p. 53.]

5. Using the definition of $\tau(m)$, we have the equation

$$\sum_{p \le x} \tau(p-1) = 2 \sum_{h \le \sqrt{x}} \pi(x; k, 1) + O\left(\sum_{\substack{p-1 = nm \le x \\ n \le \sqrt{x}, m \le \sqrt{x}}} 1\right).$$

The sum over k can be divided into two parts: For $k \le \sqrt{x}(\ln x)^{-B}$ we apply exercise 6 of Chapter IX and 8 of Chapter V; for $\sqrt{x}(\ln x)^{-B} < k \le \sqrt{x}$ we use the previous exercise. We estimate the sum σ inside the O sign for $n \le \sqrt{x}(\ln x)^{-B}$ as follows:

$$\sigma \le \sum_{r \le x(\ln x)^{-B}} \tau'(r), \text{ where } \tau'(r) \le \tau(r);$$

$$\sigma^2 \le x(\ln x)^{-B} \sum_{r \le x(\ln x)^{-B}} \tau^2(r) \ll x^2(\ln x)^{-2B+3}.$$

(Cf. Prachar [8], p. 477.)

6. a) We can take any $\gamma > 1$. Consider the sum V

$$V = \sum_{n < p^\gamma} \Lambda(n)\left(\frac{n+k}{p}\right),$$

and apply exercise 8 of Chapter III with $N = p^\gamma$ and $u = \sqrt{N}$. We obtain

$$V = V_1 - V_2 + O(\sqrt{N} \ln N),$$

where

$$V_1 = \sum_{d \le u} \mu(d) \sum_{l \le Nd^{-1}} (\log l)\left(\frac{ld+k}{p}\right),$$

$$V_2 = \sum_{d \le u} \mu(d) \sum_{n \le u} \Lambda(n) \sum_{r \le N(dn)^{-1}} \left(\frac{ndr+k}{p}\right)$$

$$= \sum_{rn \le u} \Lambda(n) \sum_{d \le u} \mu(d)\left(\frac{rnd+k}{p}\right) + \sum_{u < rn \le N} \Lambda(n) \sum_{d \le N(rn)^{-1}} \mu(d)\left(\frac{rnd+k}{p}\right).$$

We divide the last sum into two parts, corresponding to $r \le u$ and $u < r \le N$. To each of the four sums that are created, we apply exercise 9 of Chapter VIII. The transition back to σ is accomplished by means of partial summation. Part b) follows from part a). (Cf. A.A. Karatsuba, Sums of characters with prime numbers, Izv. Akad. Nauk SSSR Ser. Mat. 34 (1970), 299–321.)

7. a) and b) are obtained just like a) and b) in the preceding exercise.

Chapter XI

1. For any y, $1 \le y \le p - 1$, if we multiply both sides of the congruence by y^m, $m = n_1 \ldots n_k$, then what we get is equivalent to

$$x_1^{n_1} + \ldots + x_k^{n_k} \equiv \lambda y^m \pmod{p},$$

i.e.

$$T = \frac{1}{p-1} \sum_{a=0}^{p-1} \sum_{x_1=0}^{p-1} \cdots \sum_{x_k=0}^{p-1} \sum_{y=1}^{p-1} e^{2\pi i \frac{a(x_1^n + \ldots + x_k^n - \lambda y^m)}{p}}$$

$$= p^{k-1} + O(p^{0,5(k-1)}).$$

2. Let $W(N_1)$ denote the number of solutions of the congruence

$$x_1^n + \ldots + x_t^n \equiv N_1 (\mathrm{mod}\, Q), \; 1 \leq x_1, \ldots, x_t \leq P.$$

Then

$$W(N_1) = Q^{-1} \sum_{a=1}^{Q} S'(a) e^{-2\pi i \frac{aN_1}{Q}} = W_0(N_1) + W_1(N_1).$$

Let $W_0(N_1)$ consists of all summands in which a is of the form

$$a = p^{\alpha-\nu} a_1, \text{ where } (a_1, p) = 1 \text{ for } 0 \leq \nu \leq \alpha/m,$$

and let $W_1(N_1)$ consist of all of the remaining summands. Use the results of Lemma 1 to prove that

$$W_0(N_1) \gg P^t Q^{-1}$$

for any t in the interval $3 \leq t \leq n-1$ and any N_1.

Next, consider the congruence

$$x_1^n + \ldots + x_t^n + u_1 + u_2 + u_0 v^n \equiv N (\mathrm{mod}\, Q),$$

$$1 \leq x_1, \ldots, x_t \leq P,$$

where u_1, u_2 are of the form u,

$$u = \xi_1^n + (p^{2r}\xi_2)^n + \ldots + (p^{2rm_1}\xi_{m_1+1})^n,$$

and where $m_1 = [m]$, r is the integer defined by the condition

$$2rn \leq \alpha/m < (2r+1)n,$$

and ξ_ν runs through all positive integers that are not multiples of p, are less than $Pp^{-2r(\nu-1)}$, and are such that ξ_ν^n are pairwise incongruent modulo p^{2rn}, where $\nu = 1, 2, \ldots, m_1 + 1$. The number u_0 has the form

$$u_0 = (\zeta_1)^n + (p^r\zeta_2)^n + \ldots + (p^{rm_1}\zeta_{m_1+1})^n,$$

where ζ_ν runs through all positive integers that are not multiples of p, are less than $\sqrt{P} p^{-r(\nu-1)}$, and are such that ζ_ν^n are pairwise incongruent modulo p^{rn}, where ν takes on all integer values that are less than $\sqrt{P}$ and not multiples of p.

Let U, U_0, and V be the sets of the integers u, u_0, and v, respectively, and let $W(N)$ be the number of solutions of our congruence. We have

$$W(N) = W_0(N) + W_1(N),$$

where

$$W_0(N) \gg p^t Q^{-1} U^2 U_0 V.$$

Next,

$$W_1(N) \ll p^t \max_{a \in W_1} \left| \sum_{u_0} \sum_{v} e^{2\pi i \frac{a u_0 v^n}{Q}} \right| Q^{-1} \sum_{a=0}^{Q-1} \left| \sum_{u} e^{2\pi i \frac{au}{Q}} \right|^2.$$

The first double sum can be estimated by using the definition of W_1 and Cauchy's inequality:

$$\left| \sum_{u_0} \sum_{v} e^{2\pi i \frac{a u_0 v^n}{Q}} \right|^2 \le U_0 \sum_{y=0}^{q-1} \eta(y) \left| \sum_{v} e^{2\pi i \frac{a_1 y v^n}{q}} \right|^2,$$

where $(a_1, q) = 1$ and $\eta(y)$ is the number of solutions of the congruence $u_0 \equiv y \pmod{q}$. The second double sum is equal to the number of solutions of the congruence $u_1 \equiv u_2 \pmod{Q}$. A calculation concludes the proof of the exercise. (Cf. A.A. Karatsuba, Waring's problem for a congruence modulo the power of a prime, Vestnik Moskov. Univ. Ser. I Mat. Mekh. 1962, no. 4, 28–38.)

3. First

$$T \le P^n T_1,$$

where T_1 is the number of solutions of the system of congruences

$$\begin{cases} x_1 + \ldots + x_n \equiv \lambda_1 \pmod{p}, \\ x_1^2 + \ldots + x_n^2 \equiv \lambda_2 \pmod{p^2}, \\ \cdots\cdots\cdots\cdots\cdots\cdots\cdots\cdots \\ x_1^n + \ldots + x_n^n \equiv \lambda_n \pmod{p^n} \end{cases}$$

$$1 \le x_1, \ldots, x_n \le p^r; \quad x_i \not\equiv x_j \pmod{p}, \quad i \ne j.$$

for some fixed set of numbers $\lambda_1, \ldots, \lambda_n$. For $t = 1, 2, \ldots, n$, we represent x_t in the form

$$x_t = x_{1,t} + p x_{2,t} + \ldots + p^{r-1} x_{r,t}.$$

In order that $x_1, \ldots, x_n$ satisfy the system, it is necessary that the variables $x_{1,1}, \ldots, x_{1,n}$ satisfy the system of congruences

$$x_{1,1}^v + \ldots + x_{1,n}^v \equiv \lambda_v \pmod{p}, \quad v = 1, \ldots, n,$$

and that, for $s = 2, \ldots, r$, the variables $x_{1,s}, \ldots, x_{1,n}$ satisfy an appropriate system of linear congruences (for fixed $x_{1,1}, \ldots, x_{1,n}$).

$$x_{1,s}(v x_{1,1}^{v-1}) + \ldots + x_{n,s}(v x_{1,n}^{v-1}) \equiv \lambda_{v,s} \pmod{p}, \quad v = s, \ldots, r,$$

where $\lambda_{s,s}, \ldots, \lambda_{r,s}$ are integers.

The number of solutions of the first system does not exceed $n!$, since it follows from the elementary theory of symmetric functions that for $p > n$ and fixed λ_v every solution of this system is a permutation of single solution. The matrix of coefficients of each of these linear systems of congruences has maximum rank because the variables x_j are incongruent modulo p. Therefore, the number of solutions exceeds p^s. For T' and T we obtain the estimates

$$T' \le n! \cdot p \cdot p^2 \ldots p^{r-1} = n! p^{\frac{r(r-1)}{2}}; \quad T \le n! p^{\frac{r(r-1)}{2}} p^n.$$

4. For $x < 16$ the statement can be verified directly. Let $x \geq 16$. We have

$$\ln(k!) = \sum_{t \leq k} \ln t = \sum_{u \leq k} \sum_{d \mid t} \Lambda(d) = \sum_{u \leq k} \psi\left(\frac{k}{u}\right); \quad \psi\left(\frac{2m}{m}\right) = \frac{(2m)!}{(m!)^2};$$

$$A = \ln\binom{2m}{m} = \psi(2m) - \psi\left(\frac{2m}{2}\right) + \psi\left(\frac{2m}{3}\right) - \cdots + \psi\left(\frac{2m}{2m-1}\right).$$

It follows from this that

$$\psi(2m) - \psi(m) \leq A \leq \psi(2m) - \psi(m) + \psi\left(\frac{2}{3}m\right).$$

Using the fact that

$$\frac{1}{\sqrt{4m}} \leq \frac{1}{4^m}\binom{2m}{m} \leq \frac{1}{\sqrt{2m+1}},$$

we can use mathematical induction to prove the inequality

$$\psi(x) < x \ln 4.$$

Next, prove that

$$\sum_{m < p \leq 2m} \ln p \geq \frac{m}{3} \ln 4 - \ln\sqrt{4m} - \sqrt{2m} \ln 4;$$

$$\sum_{m < p \leq 2m} 1 \geq \frac{m \ln 4}{3 \ln 2m} - 1 - \frac{\sqrt{2m}}{\ln 2m} \ln 4 \geq \sqrt{\frac{m}{2}}.$$

The exercise now follows from the last inequality.

5. We estimate the number of sets $(x_1, \ldots, x_k)$. Let p_s be one of the numbers $p_1, \ldots, p_n$. For each set $(x_1, \ldots, x_k)$ we consider the set $(x_1^{(s)}, \ldots, x_k^{(s)})$ that consists of the remainders after division by p_s of the numbers $x_1, \ldots, x_k$:

$$x_i \equiv x_i^{(s)} \pmod{p_s}, \quad 0 \leq x_i^{(s)} < p_s, \quad i = 1, \ldots, k.$$

The number of all sets obtained in this way for fixed p_s does not exceed

$$\binom{p_s}{n-1}(n-1)^k.$$

Indeed, for each $\bar{x} = (x_1, \ldots, x_k)$ we obtain the system of congruences

$$\bar{x} \equiv \bar{x}^{(s)} \pmod{p_s}, \quad \bar{x}^{(s)} = (x_1^{(s)}, \ldots, x_k^{(s)}), \quad s = 1, \ldots, n.$$

This system of congruences can be transformed into one congruence of the form

$$\bar{x} \equiv \bar{M} \pmod{p_1 \ldots p_n},$$

where $\bar{M} = (M_1, \ldots, M_k)$ is a fixed set of integers, and $0 \leq M_i < p_1 \ldots p_n$ for $i = 1, \ldots, k$. Since each coordinate of $\bar{x}$ does not exceed $P < p_1 \ldots p_n$, then the last congruence is equivalent to the equation $\bar{x} = \bar{M}$, i.e. the number of sets

$(x_1, \ldots, x_k) = \bar{x}$ does not exceed

$$\binom{p_1}{n-1}(n-1)^k \ldots \binom{p_n}{n-1}(n-1)^k.$$

The number of sets $(y_1, \ldots, y_k)$ that satisfy the system of equations of the exercise does not exceed $n! P^{k-n}$.

It follows from this that

$$J_2 \le (n-1)^{kn}\binom{p_1}{n-1} \ldots \binom{p_n}{n-1} n! P^{k-n} \le n^{2kn} P^{k-1}.$$

6. The inequality of the exercise is trivial for $P \le (4n^2)^n$. By exercise 4, the interval $[P^{1/n}, 2P^{1/n}]$ contains n different prime numbers, say, $p_1, \ldots, p_n$. Setting $f(x) = \alpha_1 x + \ldots + \alpha_n x^n$, we have

$$J = J(P; n, k)$$

$$= \int_0^1 \ldots \int_0^1 \left| \sum_{x_1 \le P} \ldots \sum_{x_k \le P} e^{2\pi i(f(x_1) + \ldots + j(x_k))} \right|^2 d\alpha_1 \ldots d\alpha_n.$$

We divide each set $\bar{x} = (x_1, \ldots, x_k)$ into two classes A and B: The set $\bar{x} = (x_1, \ldots, x_k)$ belongs to class A if, among the numbers $p_1, \ldots, p_n$ there exists a number p_s such that, among the numbers $x_1, \ldots, x_k$ we can find at least n numbers that are pairwise incongruent modulo p_s. All of the remaining sets belong to class B. We obtain

$$J = \int_\Omega \left| \sum_{\bar{x} \in A} + \sum_{\bar{x} \in B} \right|^2 d\Omega \le 2J_1 + 2J_2,$$

where

$$J_1 = \int_\Omega \left| \sum_{\bar{x} \in A} \right|^2 d\Omega, \quad J_2 = \int_\Omega \left| \sum_{\bar{x} \in B} \right|^2 d\Omega.$$

The integral J_2 was estimated in exercise 5. Let us estimate J_2. The number J_1 is the number of solutions of the system of equations in exercise 5 under the condition that

$$\bar{x} = (x_1, \ldots, x_k) \in A, \quad \bar{y} = (y_1, \ldots, y_k) \in A.$$

The sets $\bar{x} \in A$ can be divided into n classes, where each class corresponds to $p_s = p$ for $s = 1, \ldots n$. We obtain

$$J = \int_\Omega \left| \sum_{s=1}^n \sum_{\bar{x} \in A_s} \right|^2 d\Omega \le n \sum_{s=1}^n J_{1,s},$$

where

$$J_{1,s} = \int_\Omega \left| \sum_{\bar{x} \in A_s} \right|^2 d\Omega.$$

Furthermore,

$$J_{1,s} \leq \binom{k}{n}^2 \int_\Omega \left| \underset{x_1, \ldots, x_n}{\sum}' \right|^2 \left| \sum_{x \leq P} e^{2\pi f/(x)} \right|^{2k-2n} d\Omega,$$

where the prime $'$ in the first sum denotes summation over those sets $x_1, \ldots, x_n$ which are pairwise incongruent modulo $p_s = p$. Dividing the range of summation of the second sum into p progressions with difference p and applying Hölder's inequality, we obtain

$$J_{1,s} \leq \binom{k}{n}^2 p^{2k-2n-1} \sum_{y=1}^p \int_\Omega \left| \underset{x_1, \ldots, x_n}{\sum}' \right|^2 \left| \sum_{0 \leq z \leq Pp^{-1}} e^{2\pi i f(y+pz)} \right|^{2k-2n} d\Omega,$$

$$\leq \binom{k}{n}^2 p^{2k-2n} \int_\Omega \left| \underset{x_1, \ldots, x_n}{\sum}'' \right|^2 \left| \sum_{0 \leq z \leq Pp^{-1}} e^{2\pi i f(pz)} \right|^{2k-2n} d\Omega,$$

where the symbol Σ'' denotes summation over all sets of numbers $x_1, \ldots, x_n$ which vary between the limits $-y_0$ and $P - y_0$ and are pairwise incongruent modulo p. The last integral does not exceed

$$J(P_1; n, k - n)T,$$

where $P_1 = Pp^{-1} + 1$, and T is the number of solutions of the system of congruences in exercise 3. The inequality of the exercise follows easily from this.

7. Carry out a proof by induction in the parameter, and use the inequality of exercise 6.

(In connection with exercises 3–7, cf. A.A. Karatsuba, On systems of congruences, Izv. Akad. Nauk SSSR Ser. Mat. 29 (1965), 935–944; A.A. Karatsuba, Theorems on the mean and complete trigonometric sums, Izv. Akad. Nauk SSSR Ser. Mat. 30 (1966), 183–206; and G.I. Arkhipov, et al. [1], pp 12–29.)

8. Consider the polynomial $F(x)$,

$$F(x) = \frac{x}{1!} - 2\frac{x(x-1)}{2!} + \ldots \pm 2^{n-1}\frac{x(x-1)\ldots(x-n+1)}{n!}$$

$$\equiv 2^{-1}(1 + (1 - 2)^{x+1})(\mathrm{mod}\ 2^n).$$

It follows from the definition of $F(x)$ that

$$F(x) \equiv 0\,(\mathrm{mod}\ 2^n), \quad \text{if } x \equiv 0\,(\mathrm{mod}\ 2);$$

$$F(x) \equiv 1\,(\mathrm{mod}\ 2^n), \quad \text{if } x \equiv 1\,(\mathrm{mod}\ 2).$$

Since the greatest power of 2 that divides $n!$ is at most $n - 1$, it follows that the denominators of all of the coefficients of the polynomial $F(x)$ are odd. Therefore, there exists a polynomial $f(x)$ with integer coefficiets $a_n, \ldots, a_2, a_1 = 1$ such that for all x

$$f(x) \equiv F(x)\ (\mathrm{mod}\ 2^n).$$

9. Multiplying the first, second, ..., last congruences by $a_n, a_{n-1}, \ldots, a_1$ and adding, we obtain

$$f(x_1) + \ldots + f(x_k) \equiv a_n N_n + \ldots + a_1 N_1 \equiv l(\mathrm{mod}\ 2^n),\quad 0 \le l < 2^n.$$

It follows from the properties of $f(x)$ that

$$l \equiv k_1(\mathrm{mod}\ 2^n),$$

where k_1 is the number of odd unknowns among $x_1, \ldots, x_k$. This implies that $k \ge k_1 \ge 1$.

(In connection with exercises 8–9, cf. G.I. Arknipov, On the value of the singular series in the Hilbert-Kamke problem. Dokl. Akad. Nauk SSSR 259 (1981), 265–267; G.I. Arkhipov, On the Hilbert-Kamke problem, Izv. Akad. Nauk SSSR Ser. Mat. 48 (1984), 3–52.)

10. If x is an odd integer, then for $n = 2^k$ we have the congruence

$$x^n \equiv 1\ (\mathrm{mod}\ 4n).$$

The statement of the exercise follows from this.

11. Without loss of generality we can assume that each of the numbers $x_1, \ldots, x_k$ is odd. For $j = 1, \ldots, k$, we represent each x_j in the form

$$x_j \equiv \pm 5^{\alpha_j}(\mathrm{mod}\ 2^n),$$

and we define the polynomial $f(t)$ by the equation

$$f(t) = t^{\alpha_1} + \ldots + t^{\alpha_k}.$$

We shall prove that

$$k \equiv f(1) \equiv 0\ (\mathrm{mod}\ 2^u).$$

It follows from the definition of $f(t)$ that

$$f(5^{2j_v}) \equiv 0(\mathrm{mod}\ 2^{2j_v}),\quad v = 1, 2, \ldots, m.$$

We represent $f(t)$ in the form

$$\begin{aligned}
f(t) = a_0 &+ a_1(t - 5^{2j_m}) + a_2(t - 5^{2j_m})(t - 5^{2j_{m-1}}) + \ldots \\
&+ a_{m-1}(t - 5^{2j_m})(t - 5^{2j_{m-1}}) \ldots (t - 5^{2j_2}) \\
&+ g(t)(t - 5^{2j_m})(t - 5^{2j_{m-1}}) \ldots (t - 5^{2j_2})(t - 5^{2j_1})
\end{aligned}$$

where $a_0, a_1, a_2, \ldots, a_{m-1}$ are integers, $a_0 = f(5^{2j_m})$, and a_1 is the remainder when $f(5^{2j_{m-1}}) - a_0$ is divided by $5^{2j_{m-1}} - 5^{2j_m}$, etc., and $g(t)$ is a polynomial with integer coefficients. Furthermore,

$$\begin{aligned}
\varphi(t) = a_0 &+ a_1(t - 5^{2j_m}) + a_2(t - 5^{2j_m})(t - 5^{2j_{m-1}}) + \ldots \\
&+ a_{m-1}(t - 5^{2j_m})(t - 5^{2j_{m-1}}) \ldots (t - 5^{2j_2}) \\
&= \sum_{s=1}^{m} c_s \frac{(t - t_1) \ldots (t - t_{s-1})(t - t_{s+1}) \ldots (t - t_m)}{(t_s - t_1) \ldots (t_s - t_{s-1})(t_s - t_{s+1}) \ldots (t_s - t_m)},
\end{aligned}$$

where $t_s = 5^{2j_s}$, $c_s = \varphi(t_s) \equiv 0\ (\mathrm{mod}\ 2^{2j_s})$, $s = 1, 2, \ldots, m$.

If we denote by $\delta(v)$ the highest power of 2 that divides v, then

$$\delta(5^a - 1) = \delta(a) + 2, \quad \delta(a!) = \left[\frac{a}{2}\right] + \left[\frac{a}{2^2}\right] + \ldots$$

and so

$$\delta\left(c_s \frac{(1 - t_1)\ldots(1 - t_{s-1})(1 - t_{s+1})\ldots(1 - t_m)}{(t_s - t_1)\ldots(t_s - t_{s-1})(t_s - t_{s+1})\ldots(t_s - t_m)}\right)$$

$$\geq 2j_s - \delta(j_1 - j_s) - \ldots - \delta(j_{s-1} - j_s) - \delta(j_{s+1} - j_s) - \ldots - \delta(j_m - j_s)$$

$$\geq 2j_s - \delta((j_s - j_1)!) - \delta((j_m - j_s)!) \geq 2j_s + j_1 - j_m \geq n/32.$$

Moreover,

$$\delta(g(1)(1 - 5^{2j_m})(1 - 5^{2j_{m-1}})\ldots(1 - 5^{2j_1})) \geq 3m \geq n/32.$$

12. We shall call a form *special* if it only represents zero trivially. Prove that if F is a special form, then for any $m \geq 1$ there exists an N such that the congruence

$$F = F(x_1, \ldots, x_k) \equiv 0 \,(\text{mod } 2^N)$$

implies that

$$x_1 \equiv \ldots \equiv x_k \equiv 0 \,(\text{mod } 2^m).$$

For any natural number k, let $\kappa(k)$ equal the highest degree n such that there exists a special form F of degree n in k variables. Exercise 10 implies the inequality

$$\varkappa(k) \leq \tfrac{1}{4}(k + 1).$$

Next, prove that $\kappa(k) \leq \kappa(k + 1)$.

Now let $n = 4t$, where t is a natural number, and let $F(y_0, \ldots, y_{t-1})$ be a special form of degree $\kappa(t)$,

$$G(x_1, \ldots, x_k) = F(y_0, y_1, \ldots, y_{t-1}),$$

where

$$y_j = s_{2j} \cdot s_{n-2j}, \quad j = 0, 1, \ldots, t - 1,$$

$$S_v = x_1^v + \ldots + x_k^v, \quad v = 1, \ldots, n; \quad s_0 = 1.$$

The degree of the form G equals $n\kappa(t)$. Since the form F is special, then for some N the congruence

$$F(y_0, y_1, \ldots, y_{t-1}) \equiv 0 \,(\text{mod } 2^N)$$

implies that

$$y_0 \equiv y_1 \equiv \ldots \equiv y_{t-1} \equiv 0 \,(\text{mod } 2^{2n}),$$

i.e. we can find integers $1 \leq j_1 \leq \ldots < j_t = 2t$ such that

$$s_{2j_1} \equiv \ldots \equiv s_{2j_t} \equiv 0 \,(\text{mod } 2^n).$$

It follows from exercise 11 that $\kappa(k) \leq n\kappa(t)$ if only $k < 2^u$, $u = n/32$. From this, we can prove by induction the inequality

$$\varkappa(k) < \underbrace{(\log_2 k)(\log_2 \log_2 k) \ldots (\log_2 \ldots \log_2 k)}_{r+1} \underbrace{(\log_2 \ldots \log_2^3 k)}_{r+2},$$

and the assertion of the exercise follows from this.

13. This is solved in the same way that exercises 11 and 12 were solved.

(In connection with exercises 11–13, cf. G.I. Arkhipov and A.A. Karatsuba, Local representation of zero by a form, Izv. Akad. Nauk SSSR Ser. Mat. 45 (1981), 948–961; G.I. Arkhipov and A.A. Karatsuba, On a problem in the theory of congruences, Uspekhi Mat. Nauk 37 (1982), 161–162.)

14. Repeat the proof of Theorem 2 of Chapter XI. (Cf. Vinogradov [10]; A.A. Karatsuba, The mean value of the modulus of a trigonometric sum, Izv. Akad. Nauk SSSR Ser. Mat. 37 (1973), 1203–1227.)

15. a) Repeat the solution of exercise 2 of Chapter X. Estimate the integral I_2 by means of exercise 14. The main term has the form

$$\sum_{x < P} I(N - x^n)\sigma(N - x^n), \quad P = \sqrt[n]{N},$$

where

$$I(N - x^n) = \sum_{m + m_1 = N - x^n} \frac{1}{\log m \log m_1},$$

$$\sigma(N - x^n) = \prod_{p \times N - x^n} \left(1 - \frac{1}{(p-1)^2}\right) \prod_{p \backslash N - x^n} \left(1 + \frac{1}{p-1}\right).$$

(Cf. T. Estermann, Proof that every large integer is the sum of two primes and a square, Proc. London Math. Soc. 11 (1937), 501–516.)

b) Repeat the solution to a), and use the results of section 1 of Chapter XI.

Table of Prime Numbers <4070 and their Smallest Primitive Roots

p	g	p	g	p	g	p	g	p	g	p	g	p	g
2	1	151	6	353	3	577	5	811	3	1049	3	1297	10
3	2	157	5	359	7	587	2	821	2	1051	7	1301	2
5	2	163	2	367	6	593	3	823	3	1061	2	1303	6
7	3	167	5	373	2	599	7	827	2	1063	3	1307	2
11	2	173	2	379	2	601	7	829	2	1069	6	1319	13
13	2	179	2	383	5	607	3	839	11	1087	3	1321	13
17	3	181	2	389	2	613	2	853	2	1091	2	1327	3
19	2	191	19	397	5	617	3	857	3	1093	5	1361	3
23	5	193	5	401	3	619	2	859	2	1097	3	1367	5
29	2	197	2	409	21	631	3	863	5	1103	5	1373	2
31	3	199	3	419	2	641	3	877	2	1109	2	1381	2
37	2	211	2	421	2	643	11	881	3	1117	2	1399	13
41	6	223	3	431	7	647	5	883	2	1123	2	1409	3
43	3	227	2	433	5	653	2	887	5	1129	11	1423	3
47	5	229	6	439	15	659	2	907	2	1151	17	1427	2
53	2	233	3	443	2	661	2	911	17	1153	5	1429	6
59	2	239	7	449	3	673	5	919	7	1163	5	1433	3
61	2	241	7	457	13	677	2	929	3	1171	2	1439	7
67	2	251	6	461	2	683	5	937	5	1181	7	1447	3
71	7	257	3	463	3	691	3	941	2	1187	2	1451	2
73	5	263	5	467	2	701	2	947	2	1193	3	1453	2
79	3	269	2	479	13	709	2	953	3	1201	11	1459	5
83	2	271	6	487	3	719	11	967	5	1213	2	1471	6
89	3	277	5	491	2	727	5	971	6	1217	3	1481	3
97	5	281	3	499	7	733	6	977	3	1223	5	1483	2
101	2	283	3	503	5	739	3	983	5	1229	2	1487	5
103	5	293	2	509	2	743	5	991	6	1231	3	1489	14
107	2	307	5	521	3	751	3	997	7	1237	2	1493	2
109	6	311	17	523	2	757	2	1009	11	1249	7	1499	2
113	3	313	10	541	2	761	6	1013	3	1259	2	1511	11
127	3	317	2	547	2	769	11	1019	2	1277	2	1523	2
131	2	331	3	557	2	773	2	1021	10	1279	3	1531	2
137	3	337	10	563	2	787	2	1031	14	1283	2	1543	5
139	2	347	2	569	3	797	2	1033	5	1289	6	1549	2
149	2	349	2	571	3	809	3	1039	3	1291	2	1553	3

p	g	p	g	p	g	p	g	p	g	p	g	p	g
1559	19	1901	2	2267	2	2621	2	2957	2	3343	5	3697	5
1567	3	1907	2	2269	2	2633	3	2963	2	3347	2	3701	2
1571	2	1913	3	2273	3	2647	3	2969	3	3359	11	3709	2
1579	3	1931	2	2281	7	2657	3	2971	10	3361	22	3719	7
1583	5	1933	5	2287	19	2659	2	2999	17	3371	2	3727	3
1597	11	1949	2	2293	2	2663	5	3001	14	3373	5	3733	2
1601	3	1951	3	2297	5	2671	7	3011	2	3389	3	3739	7
1607	5	1973	2	2309	2	2677	2	3019	2	3391	3	3761	3
1609	7	1979	2	2311	3	2683	2	3023	5	3407	5	3767	5
1613	3	1987	2	2333	2	2687	5	3037	2	3413	2	3769	7
1619	2	1993	5	2339	2	2689	19	3041	3	3433	5	3779	2
1621	2	1997	2	2341	7	2693	2	3049	11	3449	3	3793	5
1627	3	1999	3	2347	3	2699	2	3061	6	3457	7	3797	2
1637	2	2003	5	2351	13	2707	2	3067	2	3461	2	3803	2
1657	11	2011	3	2357	2	2711	7	3079	6	3463	3	3821	3
1663	3	2017	5	2371	2	2713	5	3083	2	3467	2	3823	3
1667	2	2027	2	2377	5	2719	3	3089	3	3469	2	3833	3
1669	2	2029	2	2381	3	2729	3	3109	6	3491	2	3847	5
1693	2	2039	7	2383	5	2731	3	3119	7	3499	2	3851	2
1697	3	2053	2	2389	2	2741	2	3121	7	3511	7	3853	2
1699	3	2063	5	2393	3	2749	6	3137	3	3517	2	3863	5
1709	3	2069	2	2399	11	2753	3	3163	3	3527	5	3877	2
1721	3	2081	3	2411	6	2767	3	3167	5	3529	17	3881	13
1723	3	2083	2	2417	3	2777	3	3169	7	3533	2	3889	11
1733	2	2087	5	2423	5	2789	2	3181	7	3539	2	3907	2
1741	2	2089	7	2437	2	2791	6	3187	2	3541	7	3911	13
1747	2	2099	2	2441	6	2797	2	3191	11	3547	2	3917	2
1753	7	2111	7	2447	5	2801	3	3203	2	3557	2	3919	3
1759	6	2113	5	2459	2	2803	2	3209	3	3559	3	3923	2
1777	5	2129	3	2467	2	2819	2	3217	5	3571	2	3929	3
1783	10	2131	2	2473	5	2833	5	3221	10	3581	2	3931	2
1787	2	2137	10	2477	2	2837	2	3229	6	3583	3	3943	3
1789	6	2141	2	2503	3	2843	2	3251	6	3593	3	3947	2
1801	11	2143	3	2521	17	2851	2	3253	2	3607	5	3967	6
1811	6	2153	3	2531	2	2857	11	3257	3	3613	2	3989	2
1823	5	2161	23	2539	2	2861	2	3259	3	3617	3	4001	3
1831	3	2179	7	2543	5	2879	7	3271	3	3623	5	4003	2
1847	5	2203	5	2549	2	2887	5	3299	2	3631	15	4007	5
1861	2	2207	5	2551	6	2897	3	3301	6	3637	2	4013	2
1867	2	2213	2	2557	2	2903	5	3307	2	3643	2	4019	2
1871	14	2221	2	2579	2	2909	2	3313	10	3659	2	4021	2
1873	10	2237	2	2591	7	2917	5	3319	6	3671	13	4027	3
1877	2	2239	3	2593	7	2927	5	3323	2	3673	5	4049	3
1879	6	2243	2	2609	3	2939	2	3329	3	3677	2	4051	6
1889	3	2251	7	2617	5	2953	13	3331	3	3691	2	4057	5

Bibliography

1. Arkhipov, G.I., Karatsuba, A.A., Chubarikov, V.N.: Multiple trigonometric sums (Russian), Trudy Mat. Inst. Steklova 151 (1980); translation: Proc. Steklov Inst. Math. 1982, no. 2, American Mathematical Society, Providence

2. Chandrasekharan, K.: *Arithmetical Functions*, Springer-Verlag, Berlin 1970

3. Chudakov, N.G.: Introduction to the Theory of Dirichlet's L-functions, Gostekhizdat, Moscow 1947

4. Davenport, H.: *Multiplicative Number Theory*, second edition (revised by Hugh L. Montgomery), Springer-Verlag, Berlin 1980

5. Hua Lo-Keng: *Die Abschätzung von Exponentialsummen und ihre Anwendung in der Zahlentheorie*, Enzyklopädie der mathematischen Wissenschaften, I.2, Heft 13, Teil 1, Teubner, Leipzig 1959

6. Ingham, A.E.: *The Distribution of Prime Numbers*, Cambridge Mathematical Tracts No. 30, Cambridge University Press, Cambridge 1932

7. Montgomery, L.: *Topics in Multiplicitive Number Theory*, Lecture Notes in Mathematics, Vol. 227, Springer-Verlag, Berlin 1971

8. Prachar, K.: *Primzahlverteilung*, Springer-Verlag, Berlin 1957

9. Titchmarsh, E.C.: *The Theory of the Riemann Zeta-Function*, second edition (revised by D.R. Heath-Brown), Clarendon Press, Oxford 1986

10. Vinogradov, I.M.: *The Method of Trigonometric Sums in Number Theory*, second edition, translated in: Ivan Matveevich Vinogradov, *Selected Works*, Springer-Verlag, 1985, pp. 181–295

11. Vinogradov, I.M.: *Selected Works*, Soviet Academy of Sciences Press, Moscow 1952

12. Vinogradov, I.M.: *Special variants of the method of trigonometric sums*, translated in: Ivan Matveevich Vinogradov, *Selected Works*, Springer-Verlag, 1985, pp. 299–383

Subject Index